7th Symposium on the Mechanics of Slender Structures (MoSS2017)

Journal of Physics: Conference Series
Volume 1048

Merida, Spain
14 - 15 December 2017

Editors:

Ignacio Herrera Navarro
Stefan Kaczmarczyk

ISBN: 978-1-5108-6789-5
ISSN: 1742-6588

Printed from e-media with permission by:

Curran Associates, Inc.
57 Morehouse Lane
Red Hook, NY 12571

Some format issues inherent in the e-media version may also appear in this print version.

Printed by Curran Associates, Inc. (2018)

Institute of Physics
Dirac House, Temple Back
Bristol BS1 6BE UK

Phone: 44 1 17 929 7481
Fax: 44 1 17 920 0979

techtracking@iop.org

Additional copies of this publication are available from:

Curran Associates, Inc.
57 Morehouse Lane
Red Hook, NY 12571 USA
Phone: 845-758-0400
Fax: 845-758-2633
Email: curran@proceedings.com
Web: www.proceedings.com

Table of contents

Volume 1048

The 7th Symposium on Mechanics of Slender Structures

14–15 December 2017, Mérida, Spain

Accepted papers received: 21 June 2018
Published online: 4 July 2018

Preface

Papers

IOP Conf. Series: Journal of Physics: Conf. Series **1048** (2018) 011001 doi:10.1088/1742-6596/1048/1/011001

Preface

7th Symposium on Mechanics of Slender Structures 2017 (MoSS2017)

I Herrera[1] S Kaczmarczyk[2]

[1]University of Extremadura, Spain, e-mail: iherrera@unex.es
[2]The University of Northampton, UK, e-mail: stefan.kaczmarczyk@northampton.ac.uk

This issue of Journal of Physics: Conference Series contains selected technical papers that were presented at the 7th Symposium on the Mechanics of Slender Structures (MoSS2017) and show a range of different approaches used in the analysis of slender structures. The meeting was hosted by The Lift Engineering Research Center (INGELEV) of the University of Extremadura, and was held in Mérida, Spain, from 14th to 15th December 2017. This conference runs under the auspices of the Institute of Physics Applied Mechanics Group and forms a continuation of a successful meeting series on the Mechanics of Slender Structures first held in Northampton, UK, in 2006, followed by the event hosted in Baltimore, USA in 2008, in San Sebastian, Spain in 2010, in Harbin, China in 2013, in Northampton, U.K. in 2015 and in Shanghai, China in 2016.

The remit of the Symposium on the Mechanics of Slender Structures series involves a broad range of scientific areas. Applications of slender structures include terrestrial, marine and space systems. Moving elastic elements such as ropes, cables, belts and tethers are pivotal components of many engineering systems. Their lengths often vary when the system is in operation. The applications include vertical transportation installations and, more recently, space tether propulsion systems. Traction drive elevator (lift) installations employ ropes and belts of variable length as a means of suspension, and also for the compensation of tensile forces over the traction sheave. In cranes and mine hoists, cables and ropes are subject to length variation in order to carry payloads. Tethers experiencing extension and retraction are important components of offshore and marine installations, as well as being proposed for a variety of different space vehicle propulsion systems based on different applications of momentum exchange and electrodynamic interactions with planetary magnetic fields. Furthermore, cables and slender rods are used extensively in civil engineering; in cable-supported bridges, guyed masts and long-span roofs of buildings and stadia. Suspended cables are also applied as electricity transmission lines. Chains are used in various power transmission systems that include such mechanical systems as chain drives and chain saws. Moving conveyor belts are essential components in various material handling installations. Other structures such as pipelines, plates, membranes, beams, mechanical linkages, and DNA structures also fall into this category.

Tall buildings and towers represent another important class of slender structures. In the modern high-rise built environment tall buildings have increased height and slenderness as well as reduced weight. Such structures are designed to withstand a broad range of external loads such as strong wind and seismic excitation. However, they are prone to structural vibrations and complex resonance phenomena that cause damage, affect their occupants and modular installations such as vertical transportation/lift systems. The performance of these installations plays a significant role in the building operation and a holistic approach is needed in the analysis and design of the structural system.

IOP Conf. Series: Journal of Physics: Conf. Series **1048** (2018) 011001 doi:10.1088/1742-6596/1048/1/011001

The MoSS 2017 involved papers devoted to special interests in recent multi-disciplinary applications in the mechanics of slender structures, such as bones mechanics, implants, tendons, soft materials, aeroplane wings, mega-long viaducts and in other new cutting edge technologies.

This meeting brought together experts from various fields: structural mechanics, bio-mechanics, thermo-mechanics, dynamics, electrodynamics, vibration and control, structural health monitoring, artificial intelligence, materials science and applied mathematics to discuss the current state of research as well as rising trends and direction for future research in the area of mechanics of slender structures.

The event is aimed at improving the understanding of structural and thermo-mechanical properties and behaviour of slender structures. The papers presented at the conference covered analytical, numerical and experimental research in various applications of slender structures. The conference programme was arranged around the following six keynote lectures:

Keynote Lecture 1: Linear and nonlinear dynamic cable-structure interactions
by John Macdonald, University of Bristol (UK).

Keynote Lecture 2: Pros and Cons of Beams Modelled with the Absolute Nodal Coordinate Formulation
by Johannes Gerstmayr, University of Innsbruck (Austria).

Keynote Lecture 3: Present and Future Vertical Transportation Systems
by Rory Smith, thyssnkrupp Elevator (USA)/ University of Northampton (UK).

Keynote Lecture 4: Two decades of research on vibrations of continuous systems and dynamics of elevator systems
by Weidong Zhu, University of Maryland BC (USA).

Keynote Lecture 5: WYPiWYG constitutive modelling for soft materials
by Francisco Javier Montans, University of Madrid (Spain)

Keynote Lecture 6: Almonte Viaduct in the Madrid – Portuguese Border High Speed Line
by Pablo Jiménez, ADIF (Spain)

The articles presented in this volume are arranged alphabetically by the first author details and are extended peer-reviewed versions of the papers presented at the conference. The Editors and Organizing Committee gratefully acknowledges support received from the co-sponsoring institutions and would like to thank the authors for their hard work and high quality contributions.

IOP Conf. Series: Journal of Physics: Conf. Series **1048** (2018) 011001 doi:10.1088/1742-6596/1048/1/011001

Editors:

Ignacio Herrera Navarro, University of Extremadura, Spain (Co-Editor)
Stefan Kaczmarczyk, The University of Northampton, UK (Co-Editor)

The Organizing Committee:

Professor Ignacio Herrera Navarro, University of Extremadura, Spain (Co-Chair)
Professor Stefan Kaczmarczyk, University of Northampton, U.K. (Co-Chair)
Dr. Francisco Javier Alonso Sánchez, University of Extremadura, Spain
David Polo González, University of Extremadura, Spain

IOP | Institute of Physics Stress and Vibration Group

UNIVERSIDAD DE EXTREMADURA

CIBSE LIFTS GROUP

Aecae

FEEDA FEDERACIÓN EMPRESARIAL ESPAÑOLA DE ASCENSORES

14-15 December 2017, Mérida Congress Centre, Mérida, Extremadura.

Peer review statement

All papers published in this volume of *Journal of Physics: Conference Series* have been peer reviewed through processes administered by the proceedings Editors. Reviews were conducted by expert referees to the professional and scientific standards expected of a proceedings journal published by IOP Publishing.

IOP Conf. Series: Journal of Physics: Conf. Series **1048** (2018) 012001 doi:10.1088/1742-6596/1048/1/012001

Design of slender and lightweight rehabilitation orthotics based on Lines of Non Extension

J. Barrios-Muriel[1]., G. Rodríguez Jiménez[1], F. Romero Sánchez[1]. F. J. Alonso Sánchez[1]., D. Rodríguez Salgado[1].

[1]Escuela de Ingenierías Industriales. Universidad de Extremadura. Avda. de Elvas s/n, 06006 Badajoz, Spain

Email: jorgebarrios@unex.es

Abstract. The development of dynamic orthoses is a fast-growing field of research and has resulted in many different devices, where all of them try to solve or correct some motion pathology. As these devices are held on the human skin, is behaviour must be a source of inspiration for the safety design of wearable devices [1]. Furthermore, to design a comfortable orthosis the joint motion must be considered. Based on this, the approach of this work for the design of rehabilitation orthotics is to analyse the skin strain field during a prescribed rehabilitation motion. Then, based on anatomical lines with minimum deformation, also called, Lines of non-extension (LoNES) it is possible to design the structural parts of the orthosis. This LoNEs can be obtained from the recorded motion by Digital Image Correlation and different optimization criteria. The structural parts of the orthotic can be designed over these lines, and consequently these parts must be considered as slender structures. In this work, the complete process to obtain a functional orthotic device is presented. The design of the slender contours of the orthotic and the optimization (material selection, thicknes, etc.) based on finite element analysis is also described. Different prototypes are analysed to determine the areas with major strain. The resulting prototype is optimize the design based on this information.

1. Introduction

The use of the human body as support environment, and must consider the interaction with the wearable device (J.L.Pons, 2008). The relative motion and tangential strains produced in the skin-device interface during everyday life, caused by the loads due to human motion, generate wounds and the formation of bedsores and chafing. Some recommendations have been made about the use skin strain analysis as criteria of design, such as the use of the areas with relatively low movement and minimum strain for support the structure of wearable devices (2,3).
A new engineering design trend has emerged around the idea that the products in contact the human body should be on minimum strain anatomical lines, namely Lines of Non-Extension (LoNEs). Different studies have used and explored the LoNEs for different purposes, such as the development of new space suits for improving the mobility of (4–6), pattern design of tight-fit of functional clothing (Seo et al. 2013) and design of socket prostheses and orthoses (7,8). From the perspective of wearable devices, there is still no transfer of results of the skin strain field and LoNEs to the design of these products. In this sense, we have developed a new methodology of design based on skin strain analysis and LoNEs to

create wearable devices. To perform the skin strain analysis, a low-cost methodology using the principles of stereo digital image correlation (or also called 3D-DIC) has been implemented for this purpose. Based on this analysis, the basic structure, considered a slender structure, of a new design of wearable devices, such as an ankle-foot-orthosis (AFO) is presented.

2. Methods

The concept of "lines of non-extension" (LoNEs) introduced by Arthur S. Iberall (Iberall, 1964), defines lines or contours along of the human skin where there is a minimal strain during motion. Our design principle is based on the idea of analyse the skin strain field during a prescribed rehabilitation motion to obtain the LoNEs and to improve the placement of the structure of the orthosis in order to boost comfort and usability. This new methodological approach combines 3D-DIC algorithm to measure the skin deformation, LoNEs and CAD/CAE tools based on non-uniform rational B-spline (NURBS) to design the structural parts of the wearable orthosis with complex shapes due to the anatomy of the body segment.

DIC is a non-contact optical tool for surface deformation measurement that has been widely accepted and used in the field of experimental. The main advantages of this methodology are the simplicity of the data acquisition devices, the ease of measurement and also its application in large deformation cases. In essence, DIC is a technique based on digital image processing and numerical computing whose basic principle is to track the same physical points located in the undeformed and deformed images. In 3D-DIC (also called stereo-DIC), the 3D surface object is reconstructed by processing the images captured by two cam45 eras with different perspectives. A random gray-value dot pattern is painted on the surface (in this case on the human skin) to perform a correspondence between points of the two images using the DIC algorithm. A comprehensive description of the metrological aspects can be found in (9,10). The zero-mean normalized correlation criterion (ZNCC) has been used in this study as it is robust against changes in the deformed image's scale and intensity offset Pan et al. 2007).

$$\psi = \frac{\sum_a^b \sum_a^b [f(\zeta,\eta) - \bar{\bar{f}}][g(\zeta',\eta') - \bar{g}]}{\sqrt{\sum_a^b \sum_a^b [f(\zeta,\eta) - \bar{\bar{f}}]^2} \sqrt{\sum_a^b \sum_a^b [g(\zeta,\eta) - \bar{g}]^2}} \tag{1}$$

where ψ is the ZNCC, and $f(\zeta,\eta)$ are the greyscale values at the pixel $g(\zeta',\eta')$ in the reference and shift window, respectively. $\bar{\bar{f}}$ is the mean value of $f(\zeta,\eta)$ over the reference subset of dimension a x b pixels and $\bar{g}$ is the mean value of $g(\zeta',\eta')$ over the subset of current image. The maximum value of ψ is the best match between two images. Three primary steps are differentiated in the proposed methodology: stereo-system calibration, know control points. image recording and matching and, finally, 3D reconstruction and computation of the strain field (figure 3).

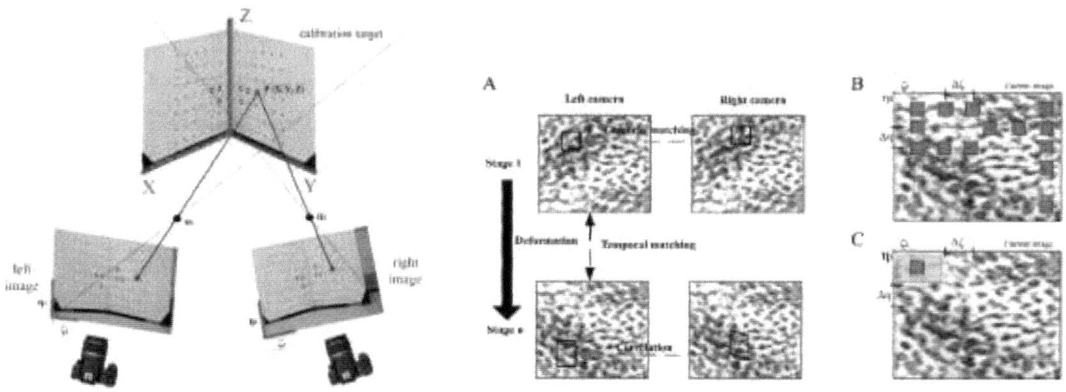

Figure 1. Left, schematic representation of the calibration process. Right, the correspondence of point for the stereo pair (A) is performed in order to determine its 3D location by triangulation. Then a deformation correspondence (B-C) is performed between the reference frame and the deformed frame to track the motion of the same point.

2.1. Experimental Setup

2.2. In this study, the skin strain field at the ankle joint was measured for one subject. Two DSLR cameras (Nikon D3300, Full HD-1080p) was used to create the stereo system and to obtain simultaneously the images of ankle-foot during the support phase of the human gait. A 22mm focal length lens with low-distortion and F-number of 14 (f/14) was used to provide a good depth of field. The setting of both camera was manually changed to obtain the best shutter speed (about 1/250) given the lighting conditions. A LED light to illuminate the ankle-foot was used. To calibrate the system, a specialized frame to calibrate stereo systems with different size of black and white pattern of dot was used. All image acquisition and 3D-DIC procedure was performed in MATLAB (Mathworks Inc.) using an-own developed toolbox. Tangential component of strain, principal strain magnitudes and LoNEs corresponding to the support phase heel contact, mid-stance and toe-foot was analysed. Both sides of the foot, both lateral and medial sides were recorded. The subject was instructed to keep a natural gait during and random step were selected to the study. Image of reference was captured standing in a static position.

3. Results and Conclusions

Tangential component of strain, principal strain magnitudes and LoNEs corresponding to the support phase heel contact, mid-stance and toe-off was analysed. Both sides of the foot, both lateral and medial sides were recorded. In figure 6 shows the tangential and principal strains values in lateral and medial view during different stages of stance phase.

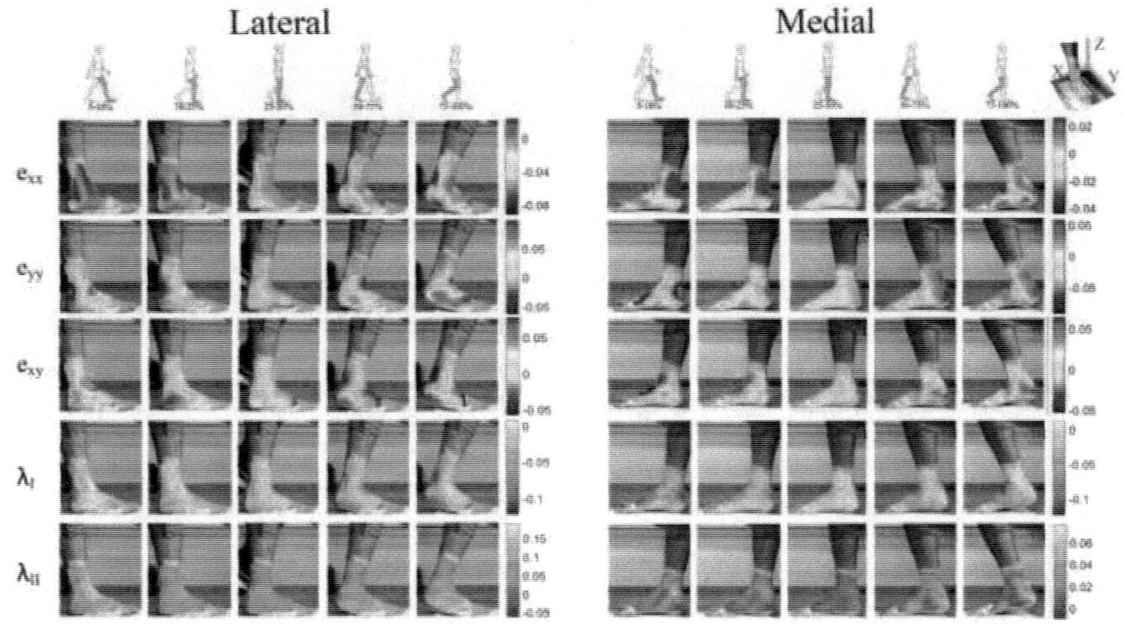

Figure 2. Skin strain field for internal and external foot. Tangential strain (e_{xx},e_{yy},e_{xy}) and principal strains magnitudes λ_I and λ_{II} (eigenvalues) are show.

IOP Conf. Series: Journal of Physics: Conf. Series **1048** (2018) 012001 doi:10.1088/1742-6596/1048/1/012001

Based on this analysis, LoNEs can be identified applying the strain ellipse theory and polar decomposition theorem. The directions of the LoNEs can be calculated by the following expression: The text of your paper should be formatted as follows:

$$\theta = \tan^{-1}\left(\sqrt{-\frac{\lambda_I}{\lambda_{II}}}\right) \tag{2}$$

where λ_I and λ_{II} are the principal strain magnitudes (eigenvalues).

Once obtained the LoNES and skin strain field, the orthoses design is created in a CAD/CAE environment. To begin the design of the virtual prototype, the triangular mesh (.obj file format) of body segment, previously obtained from the cloud points, is imported to a CAD/CAE software. Based on this mesh, section curves with B-spline can be constructed and converted in a mathematical NURBS surface (Concheiro et al. 2014; Inoue, Kikuchi, and Masuyama 2005). An offset of 2 mm is applied to the triangular mesh to avoid discomfort and to prevent the physiological swelling. From the NURBS shaped cut of the CAD surface it is transformed into IGES format for input to a CAD/CAE environment and then the virtual 3D model of wearable orthosis is constructed.

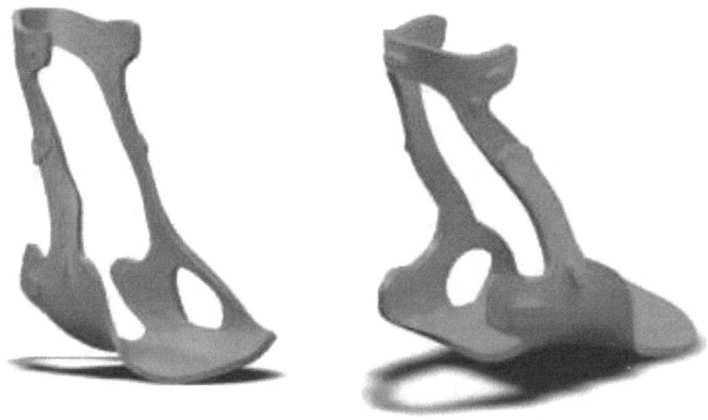

Figure 3. CAD model of orthoses design based on LoNES.

To check the structural integrity, a static analysis in a CAD/CAE environment was performed to obtain the mayor stress zones of orthosis during human gait. The direction of the loads was chosen based on the reaction force direction's vector in the stance phases of a human gait (figure 4). The material applied was ABS due to is the principal material of the convectional 3D printers and use in other work of the bibliography (Faustini et al. 2008; Papi et al. 2015).

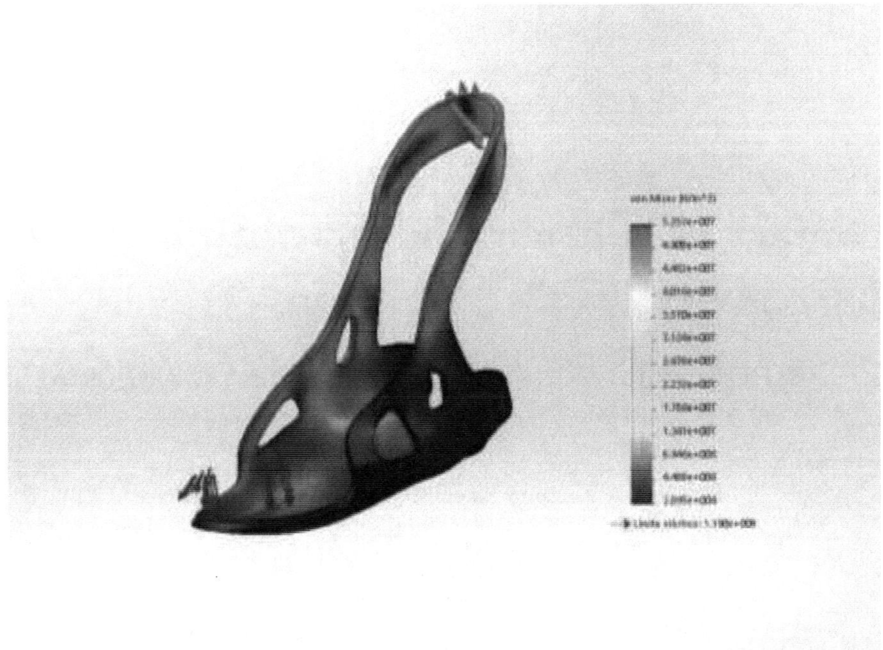

Figure 4. Static analysis of orthosis in CAD/CAE environment.

In figure 4 can be seen that the major stress areas are situated around the thinner parts connecting the sole with the calf shank. Thus, a major reinforce in these zones must be done to ensure the structural integrity of the orthoses.

References

[1] L.Pons J. Wearable Robots. Wiley; 2008. 1-361 p.

[2] Forner-Cordero A, Pons JL, Turowska EA, Schiele A, Baydal-Bertomeu JM, Garrido D, et al. Kinematics and dynamics of wearable robots. *Wearable Robot biomechatronic exoskeletons.* 2008;47–85.

[3] Mitchell RH, Medland AJ, Salo AIT. A design methodology to create constraint-based human movement patterns for ergonomic analysis. *Journal of Engineering design.* ;18(4):293–310.

[4] Newman D, Hoffman J, Bethke K, Blaya J. Astronaut bio-suit system for exploration class missions. NIAC Phase II Final. 2005;

[5] Wessendorf AM, Newman DJ. Dynamic understanding of human-skin movement and strain-field analysis. *IEEE Transactions on Biomedical Engineering.* 2012;59(12):3432–8.

[6] Obropta Jr EW. On the deformation of human skin for mechanical counter pressure space suit development. *Massachusetts Institute of Technology*; 2015.

[7] Lin B, Moerman KM, McMahan CG, Pasch KA, Herr HM. Low-Cost Methodology for Skin Strain Measurement of a Flexed Biological Limb. *IEEE Transactions on Biomedical Engineering,* 64(12), 2750-2759.

[8] Domingues Marreiros, S.P., Martins, J.M., Silva, M.T., Newman, D.J. AR. Skin Strain Field Analysis of the Human Ankle Joint. In: 4th Congresso Nacional de Biomechanica. 2010.

[9] Pan B, Xie H, Wang Z. Equivalence of digital image correlation criteria for pattern matching. *Appl Opt.* 2010; *49*(28), 5501-5509.

[10] Pan B, Xie H, Guo Z, Hua T. Full-field strain measurement using a two-dimensional Savitzky-Golay digital differentiator in digital image correlation. *Optical.* 2007; *46*(3), 033601.

Static contact of belt and pulleys with account for shear and gravity

A K Belyaev[1], **V V Eliseev**[2†], **H Irschik**[3] **and E A Oborin**[3]

[1]Institute for Problems in Mechanical Engineering, Russian Academy of Sciences, St. Petersburg, Russia

[2]Deceased on 14 October 2017. Department of Mechanical Engineering and Design, Peter the Great Saint-Petersburg Polytechnic University, St. Petersburg, Russia

[3]Institute of Technical Mechanics, Johannes Kepler University, Linz, Austria

E-mail: evgenii.oborin@jku.at

Abstract. The fitting of a looped belt on two pulleys of equal radii is considered. The belt is subject to the gravity force. A geometrically nonlinear model with account for tension and transverse shear is applied for modeling the belt. The pulleys are considered rigid bodies, and the belt-pulley contact is assumed frictionless. The problem has an axis of symmetry, therefore the boundary value problem is formulated and solved for a half of the belt. The considered part consists of three segments, two free span segments and a contact segment between them. The introduction of a dimensionless material coordinate at all segments leads to a system of ordinary differential equations. The nonlinear boundary value problem for this system and boundary conditions is solved numerically with the finite difference method. As a result, the belt shape including the rotation angle, the forces, moments and contact pressure are determined. The contact pressure increases near the end point of contact areas, however no concentrated contact forces occur. The influence of gravity force on the contact pressure is discussed.

1. Introduction

The first systematic study of belt drives (and elastic creep in it) was reported by Reynolds in [1], where he used the string model. Until recently, one-dimensional models of elastic strings were widely used, cf. [2, 3, 4]. The exact solution for nonlinear steady state equations of extensible string is obtained in [3] by assuming zones of perfect and sliding friction contact between the belt and the pulley.

A spatial description of the belt contour motion is suggested in [2]. An idealized point friction model allows extending this result to the transient dynamics, see [4].

However it turned out that the model of extensible string describes just a part of important effects in belt mechanics. Friction forces transmit power between the belt and the pulleys. They are applied on the belt from one side and result not only in tangent forces, but also in distributed moments. The model of extensible string without bending stiffness cannot resist the moment loading. Therefore, we use the rod model (i.e., the one with at least bending stiffness).

The calculation of belt-pulley interaction can be found in works [3, 5, 6, 7, 8, 9, 10]. The study of interaction should be made with account for the transverse shear, because the importance of shear in contact problems of the rod theory is well-known, cf. [11, 12, 13, 14]. A contrary point of view is presented in [15]. The introduction of shear deformation causes the absence of lumped contact forces and promotes better understanding of the contact force distribution. The shear is also required to describe the effect of elastic microslip, see, e. g. [16, 17] for general friction modeling. Particularly the shear is of importance in friction belt drive operation [18, 19, 20].

However, shear is known to be taken into account only together with tension (compression) if the nonlinear theory of rods is applied. The goal of this work is to model the belt as a rod with bending and shear stiffness in the static problem where the belt is set on the pulleys. The system of equations of the corresponding rod theory can be found in [21, 22, 23] (without direct application to belt drive mechanics, though).

The penalty formulation is in common use in numerical modeling of contact problems [24] and includes the choice of penalty coefficient or its distribution (see e. g. [25] for application to contact between rods). This approach is applied for belt drives in [10]. In the present paper we obtain the contact pressure and the stress-strain state in a different manner by describing the form of the belt lying on the rigid pulleys (the constraint formulation or the Hertz—Signorini—Moreau conditions for frictionless contact according to [24]). The similar approach of rod segmentation in contact problem can be found e. g. in [26, 27, 28], however without transverse shear.

We assume that the contact is full in a classical sense, i. e. it is continuous in an interval of the belt which length is to be determined. This assumption will be verified by obtaining the positive contact pressure. In the present study we use computer mathematics to solve difficult boundary value problems (BVP) for the systems of ordinary differential equations (ODE). The application of standard methods is possible due to introduction of transformed material coordinate.

The presented study is an extension of the work [29, 30]. Here we introduce significant complexity by including the gravity force neglected there. We intend to use these results to verify the developed finite element models [31, 32, 33].

2. Equations of nonlinear rod theory

Figure 1 displays a general calculation scheme of a rod with transverse shear and tension in the xy plane.

We consider the deformation of a belt as an elastic rod with tension and shear [21, 22, 23] subject to static contact with rigid pulley circles without friction. The position vector in the deformed configuration $\mathbf{r}(s)$ is a function of material coordinate s, and $\partial(\cdots)/\partial s = (\cdots)'$. The angular orientation of each particle of the rod is given by the orthonormal unit vectors $\mathbf{e}_1$, $\mathbf{e}_2$. The angle φ denotes the angle between unit vector $\mathbf{e}_1$ and Cartesian axis x, the latter having the unit vector $\mathbf{i}$.

For plane deformations the system of equations of nonlinear theory of rods is as follows [21, 22, 23]:

$$\mathbf{Q}' + \mathbf{q} = 0, \quad M' + \mathbf{k} \cdot \mathbf{r}' \times \mathbf{Q} + m = 0,$$
$$\varphi' - {\varphi'}_0 = AM, \quad \mathbf{r}' = \mathbf{P} \cdot {\mathbf{r}'}_0 + \mathbf{B} \cdot \mathbf{Q} = \mathbf{e}_1 (1 + B_1 Q_1) + \mathbf{e}_2 B_2 Q_2. \tag{1}$$

Here we denote: $\mathbf{Q}$ is the internal force vector, M is the bending moment, and $\mathbf{q}$, m stand for the external distributed forces and moments, respectively; $\mathbf{k}$ is the unit vector of Cartesian axis z perpendicular to the plane of drawing. The index zero indicates the values in initial state, and $\mathbf{P} = \mathbf{e}_i \mathbf{e}_{i0}$ is the rotation tensor. Elastic characteristics of the rod in plane are determined by three scalar compliances: the bending compliance A, the tension compliance B_1 and the shear compliance B_2. The unit vectors $\mathbf{e}_1$, $\mathbf{e}_2$ are directed along the principal axes of the compliance tensor $\mathbf{B}$.

Algorithms and results for the contact of belt and pulleys without shear are presented in [5, 6]. The peculiarity of the considered problem is due to the fact that the function $\mathbf{r}(s) = \mathbf{R}(\sigma)$ is unknown in the contact segment because the arc coordinate σ of the belt in its actual configuration (*solid line* in figure 2) is not yet determined.

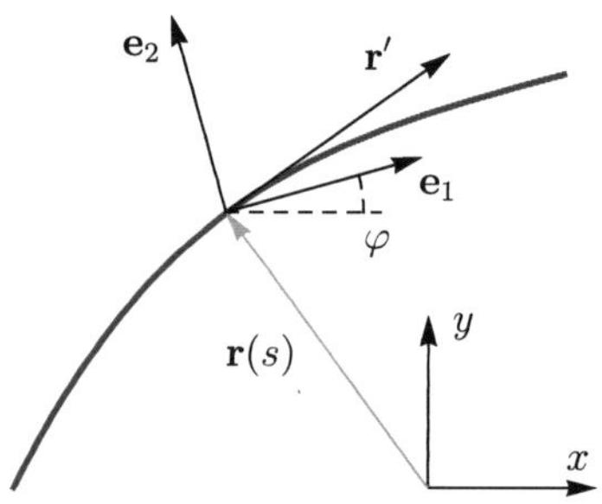

Figure 1. Cosserat rod.

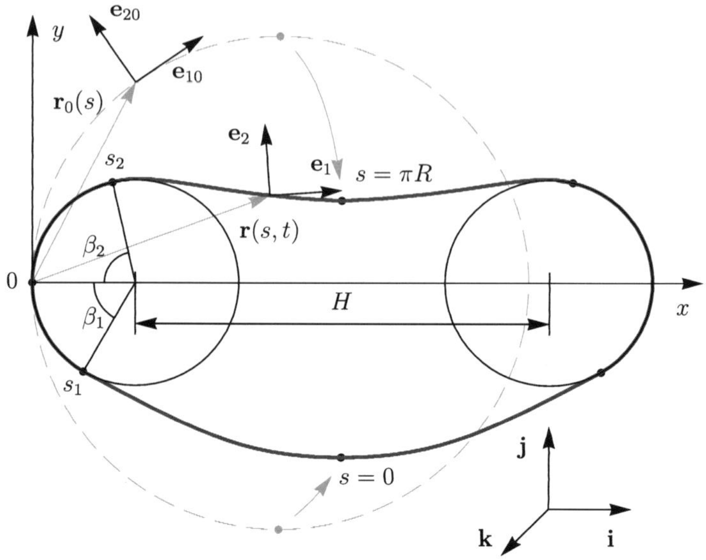

Figure 2. Belt and pulleys.

We use the following geometrical relations:

$$\mathbf{e}_1 = \mathbf{i}\cos\varphi + \mathbf{j}\sin\varphi, \quad \mathbf{e}_2 = -\mathbf{i}\sin\varphi + \mathbf{j}\cos\varphi. \tag{2}$$

We will also use the expression for $\mathbf{r}'$ from (1):

$$\mathbf{r}'(s)ds = \dot{\mathbf{R}}(\sigma)d\sigma; \ \left|\dot{\mathbf{R}}\right| = 1; \ \sigma' = |\mathbf{r}'| \equiv D = \sqrt{(1 + B_1Q_1)^2 + (B_2Q_2)^2}, \ \dot{s} = D^{-1}. \tag{3}$$

The last equations determine the relation of material coordinate s with arc coordinate σ (a dot indicates the derivative with respect to σ).

3. Contact segment

Let us consider the full contact with the pulley. We write the expression for the tangent unit vector and integrate it

$$\dot{\mathbf{R}} = \mathbf{i}\sin\kappa\sigma + \mathbf{j}\cos\kappa\sigma; \quad \mathbf{r}' = D\dot{\mathbf{R}}; \quad \mathbf{R} = \kappa^{-1}\left[\mathbf{i}\left(1 - \cos\kappa\sigma\right) + \mathbf{j}\sin\kappa\sigma\right], \tag{4}$$

where κ is the curvature of the curved belt in its actual configuration, which in the contact segment coincides with the circular pulley contour, and $\mathbf{R}(0) = 0$ is assumed. We note that

$$\mathbf{r}' \cdot \mathbf{e}_1 = 1 + B_1Q_1 = D\sin\left(\varphi + \kappa\sigma\right), \quad \mathbf{r}' \cdot \mathbf{e}_2 = B_2Q_2 = D\cos\left(\varphi + \kappa\sigma\right). \tag{5}$$

Hence we may write the force component Q_2 as a function of coordinate σ, angle φ and another force component Q_1:

$$Q_2\left(\sigma, \varphi, Q_1\right) = B_2^{-1}\left(1 + B_1Q_1\right)\cot\alpha; \quad \alpha \equiv \varphi + \kappa\sigma. \tag{6}$$

Then the relation between the derivatives (3) may be reduced to the form:

$$D = \frac{1 + B_1Q_1}{\sin\alpha}. \tag{7}$$

As a result we have four unknown functions in the contact segment which are σ, φ, M, Q_1. Let us assume no tangential contact force between the belt and the pulley. Therefore the contact pressure $\mathbf{p}$ and the tangent vector $\mathbf{r}'$ are orthogonal:

$$(\mathbf{Q}' - \rho g\,\mathbf{j}) \cdot \mathbf{r}' = 0; \;\; \mathbf{e}'_1 = \varphi'\mathbf{e}_2, \;\; \mathbf{e}'_2 = -\varphi'\mathbf{e}_1;$$
$$(Q'_1 - \varphi' Q_2 - \rho g \sin\varphi)(1 + B_1 Q_1) + (Q'_2 + \varphi' Q_1 - \rho g \cos\varphi) B_2 Q_2 = 0. \tag{8}$$

Here we may use (6) to simplify the equation (8) and write the derivatives of σ, φ Q_2

$$\varphi' = \varphi'_0 + AM,$$
$$Q'_2 = \frac{\partial Q_2}{\partial Q_1} Q'_1 + \frac{\partial Q_2}{\partial \sigma} D + \frac{\partial Q_2}{\partial \varphi} \varphi' = \frac{B_1}{B_2} Q'_1 \cot\alpha - \frac{1 + B_1 Q_1}{B_2 \sin^2\alpha} (\varphi' + \kappa\sigma'). \tag{9}$$

The combination of (8) and (9) gives the differential equation for Q_1:

$$Q'_1 = \frac{\left[\begin{array}{l} (\varphi'_0 + AM)\left(\dfrac{1 + B_1 Q_1}{B_2}\left(\dfrac{2}{1 - \cos 2\alpha} + 1\right) - Q_1\right)\cot\alpha + \\ + \dfrac{8\kappa (1 + B_1 Q_1)^2 \cos\alpha}{B_2 (\cos 4\alpha - 4\cos 2\alpha + 3)} + \rho g (\cos\varphi \cot\alpha + \sin\varphi) \end{array}\right]}{\dfrac{B_1 (1 + \cos 2\alpha)}{B_2 (1 - \cos 2\alpha)} + 1}. \tag{10}$$

Let us substitute $\mathbf{r}'$ into the balance of moments (1) (with distributed moment $m = 0$ in the current frictionless problem) and obtain

$$M' = ((B_2 - B_1) Q_1 - 1) Q_2. \tag{11}$$

Finally we determine the contact pressure p:

$$Dp = \mathbf{p} \cdot \mathbf{k} \times \mathbf{r}' = \mathbf{k} \cdot (\mathbf{Q}' - \rho g\,\mathbf{j}) \times \mathbf{r}',$$
$$p = \frac{B_2 - B_1}{B_2} Q'_1 \cos\alpha + \left(\frac{1}{B_2} + \frac{B_1 - B_2}{B_2} Q_1\right)\varphi' \sin\alpha + \frac{\kappa (1 + B_1 Q_1)^2}{B_2 \sin^2\alpha} + \rho g \sin\kappa\sigma. \tag{12}$$

The contact pressure must be non-negative. Unlike the case without shear in the considered problem there must be no concentrated contact reactions. The similar case of the classical contact between an initially straight nonlinear shearable rod and a rigid straight obstacle is discussed e. g. in [13].

4. Free segment

The scheme of fitting the belt on the pulleys is shown in figure 2. In the initial undeformed state the belt is the circle of radius R. Hence the initial angle and its derivative are $\varphi_0 = \pi - s/R$, $\varphi'_0 = -1/R$. The pulleys have curvature κ. The center distance H (i. e. the distance between the pulley centers) is prescribed. Because of the problem symmetry it is sufficient to consider only the left half $0 \le s \le \pi R$. In the segment $s_1 \le s \le s_2$ the full contact is assumed, however the coordinates s_1 and s_2 are yet unknown. The formulation for this segment is presented above.

Let us turn to the free segments (lower $0 \le s \le s_1$ and upper $s_2 \le s \le \pi R$ belt spans). Firstly we consider the lower span. The balance of forces and boundary conditions (as follows from the symmetry of the problem)

$$\mathbf{Q}' - \rho g\,\mathbf{j} = 0; \;\; Q'_x = 0, \;\; Q'_y - \rho g = 0; \;\; \mathbf{Q}(0) = -P_1\,\mathbf{i} \tag{13}$$

allow obtaining the explicit relations for the force components:

$$Q_x = -P_1 = \text{const}, \quad Q_y = \rho g s. \tag{14}$$

Here $\mathbf{Q}(0) = -P_1\mathbf{i}$ is yet unknown horizontal force between the belt left and right parts.

Then we substitute these components into the balance of moments (11), which also holds in the free segments, and here we need the relations

$$\begin{aligned} Q_1 &= Q_x \cos\varphi + Q_y \sin\varphi = -P_1 \cos\varphi + \rho g s_1 \sin\varphi, \\ Q_2 &= -Q_x \sin\varphi + Q_y \cos\varphi = P_1 \sin\varphi + \rho g s_1 \cos\varphi. \end{aligned} \tag{15}$$

The differential equations for the angle and coordinates ($\mathbf{r} = x\mathbf{i} + y\mathbf{j}$) derived from (1) are

$$\begin{aligned} \varphi' &= {\varphi'}_0 + AM, \\ x' &= \cos\varphi - P_1\left(B_1 + \frac{B_2 - B_1}{2}(1 - \cos 2\varphi)\right) + \rho g s \frac{B_1 - B_2}{2}\sin 2\varphi, \\ y' &= \sin\varphi - P_1\frac{B_1 - B_2}{2}\sin 2\varphi + \rho g s\left(B_1 + \frac{B_2 - B_1}{2}(1 + \cos 2\varphi)\right). \end{aligned} \tag{16}$$

As a result in the free segment we have four unknown functions: M, φ, x, y and an unknown constant P_1. The similar equations may be written for the second (upper) free span. The balance of forces and the boundary conditions (from the problem symmetry) allow deriving the force components

$$Q_x = P_2 = \text{const}, \quad Q_y = \rho g\,(s - \pi R). \tag{17}$$

We write the balance of moments in the form (11) where we have the force components

$$\begin{aligned} Q_1 &= P_2 \cos\varphi + \rho g(s - \pi R)\sin\varphi, \\ Q_2 &= -P_2 \sin\varphi + \rho g(s - \pi R)\cos\varphi. \end{aligned} \tag{18}$$

The remaining differential equations read

$$\begin{aligned} \varphi' &= {\varphi'}_0 + AM, \\ x' &= \cos\varphi + P_2\left(B_1 + \frac{B_2 - B_1}{2}(1 - \cos 2\varphi)\right) + \rho g(s - \pi R)\frac{B_1 - B_2}{2}\sin 2\varphi, \\ y' &= \sin\varphi + P_2\frac{B_1 - B_2}{2}\sin 2\varphi + \rho g(s - \pi R)\left(B_1 + \frac{B_2 - B_1}{2}(1 + \cos 2\varphi)\right). \end{aligned} \tag{19}$$

5. Formulation of a boundary value problem with fixed boundaries

The introduction of the new nondimensional coordinate $0 \le \xi \le 1$ (see figure 3) allows rewriting the equations in the three segments into one general BVP with fixed boundaries (see for the similar technique in general [34] and in quasi-static problem [35]). We need the relations between the derivatives:

$$\begin{aligned} 0 \le s \le s_1 &: \quad s = s_1\xi \Rightarrow {}^{d}\!/_{d\xi} = s_1 {}^{d}\!/_{ds}; \\ s_1 \le s \le s_2 &: \quad s = s_1 - \xi\,(s_1 - s_2) \Rightarrow {}^{d}\!/_{d\xi} = -\,(s_1 - s_2)\,{}^{d}\!/_{ds}; \\ s_2 \le s \le \pi R &: \quad s = \pi R - \xi\,(\pi R - s_2) \Rightarrow {}^{d}\!/_{d\xi} = -\,(\pi R - s_2)\,{}^{d}\!/_{ds}. \end{aligned} \tag{20}$$

We combine the equations at all three segments into a single system and distinguish the values by the indices $(\cdots)^{(1)}$, $(\cdots)^{(2)}$, $(\cdots)^{(3)}$, where $(\cdots)^{(1)}$ indicates the lower free segment,

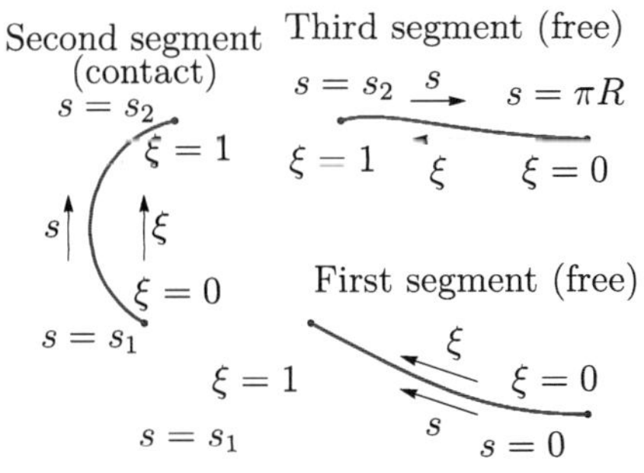

Figure 3. Segmentation of belt and coordinate transformation.

$(\cdots)^{(3)}$ the upper one, and $(\cdots)^{(2)}$ refers to the contact segment. Thus we obtain the BVP for the system of 16 first-order equations:

$$Z' = G(\xi, Z),$$
$$Z(\xi) \equiv \left(M^{(1)}\ \varphi^{(1)}\ x^{(1)}\ y^{(1)}\ \sigma^{(2)}\ Q_1^{(2)}\ M^{(2)}\ \varphi^{(2)}\ M^{(3)}\ \varphi^{(3)}\ x^{(3)}\ y^{(3)}\ s_1\ s_2\ P_1\ P_2\right)^T. \tag{21}$$

Here a prime indicates the derivative with respect to the new coordinate ξ (and the functions written in (21) are the functions of ξ). We must not forget the relations between the derivatives (20). For the unknowns $\sigma^{(2)}$, $Q_1^{(2)}$, $M^{(2)}$, $\varphi^{(2)}$ we have already derived the equations of type (3), (9), (10), (11). The remaining eight equations are the equations (11), (16), (19) and conditions for the unknown constants: $s_1' = 0$, $s_2' = 0$, $P_1' = 0$, $P_2' = 0$ (for the finite difference method the last differential equations are not necessary).

Now we consider the boundary conditions for the system (21). At $\xi = 0$ ($s = 0$ for the first segment and $s = \pi R$ for the third one, see figure 3) and $\xi = 1$ ($s = s_2$ for the second and third segments) we have

$$\begin{gathered}
\xi = 0: \quad \varphi^{(1)} = \pi,\ x^{(1)} = H/2 + \kappa^{-1},\ \varphi^{(3)} = 0,\ x^{(3)} = H/2 + \kappa^{-1}, \\
\xi = 1: \quad x^{(3)} = \kappa^{-1}\left(1 - \cos\kappa\sigma^{(2)}\right),\ y^{(3)} = \kappa^{-1}\sin\kappa\sigma^{(2)},\ \varphi^{(3)} = \varphi^{(2)}, \\
P_2\cos\varphi^{(3)} + \rho g\,(s_2 - \pi R)\sin\varphi^{(3)} = Q_1^{(2)},\ M^{(3)} = M^{(2)} \\
-P_2\sin\varphi^{(3)} + \rho g\,(s_2 - \pi R)\cos\varphi^{(3)} = B_2^{-1}\left(1 + B_1 Q_1^{(2)}\right)\cot\left(\varphi^{(3)} + \kappa\sigma^{(2)}\right).
\end{gathered} \tag{22}$$

Here we use the prescribed orientation and position (x, in other words the prescribed center distance) of the belt ends. Also we employ the fact, that the belt is continuous in x, y, φ, Q_1, Q_2, and M at the boundary points of contact zones s_1 and s_2 (no lumped forces and moments are applied there).

At the point with material coordinate s_1 we get the different values of nondimensional coordinate ξ in the first and second segment. We may write the boundary conditions in the following way:

$$\begin{gathered}
s = s_1: \quad x^{(1)}\Big|_{\xi=1} = \kappa^{-1}\left(1 - \cos\kappa\sigma^{(2)}\Big|_{\xi=0}\right),\ y^{(1)}\Big|_{\xi=1} = \kappa^{-1}\sin\kappa\sigma^{(2)}\Big|_{\xi=0}, \\
\varphi^{(1)}\Big|_{\xi=1} = \varphi^{(2)}\Big|_{\xi=0},\ -P_1\cos\varphi^{(1)}\Big|_{\xi=1} + \rho g s_1\sin\varphi^{(1)}\Big|_{\xi=1} = Q_1^{(2)}\Big|_{\xi=0},\ M^{(1)}\Big|_{\xi=1} = M^{(2)}\Big|_{\xi=0} \\
P_1\sin\varphi^{(1)}\Big|_{\xi=1} + \rho g s_1\cos\varphi^{(1)}\Big|_{\xi=1} = B_2^{-1}\left(1 + B_1 Q_1^{(2)}\Big|_{\xi=0}\right)\cot\left(\varphi^{(1)}\Big|_{\xi=1} + \kappa\sigma^{(2)}\Big|_{\xi=0}\right).
\end{gathered} \tag{23}$$

We note that to obtain the arc coordinate $\sigma(s)$ in the free segments and the whole belt length after the deformation we need to solve two ordinary differential equations (3) with the values of σ at s_1 and s_2 as initial conditions.

IOP Conf. Series: Journal of Physics: Conf. Series **1048** (2018) 012002 doi:10.1088/1742-6596/1048/1/012002

6. Numerical results

The formulated nonlinear BVP (21), (22), (23) may be solved by the shooting method. However we prefer the finite difference method because of its efficiency. We implement it in Wolfram Mathematica with the scheme described in [29, 30]. The formulated problem may be also solved by the collocation method with the solver of Matlab package Chebfun [36].

The parameters of the benchmark example are: $H = 2\,\mathrm{m}$, $\kappa = 1/0.15\,\mathrm{m}^{-1}$, $L = 1.05(H + \pi/\kappa)$, $R = L/\pi$, $E = 10^9\,\mathrm{Pa}$, $\nu = 0.5$, $\mu = E/(2(1+\nu))$, $\rho = 0.48\,\mathrm{kg/m}$ ($\rho = 0.16\,\mathrm{kg/m}$), $A = 1/0.07\,\mathrm{N}^{-1}\mathrm{m}^{-2}$, $B_1 = 1/2000\,\mathrm{N}^{-1}$, $F = 1/(EB_1)$, $B_2 = 6/(5\mu F)$, $g = 9.81\,\mathrm{m/s}^2$. We draw the belt form in figure 4 and pressure distribution in figure 5.

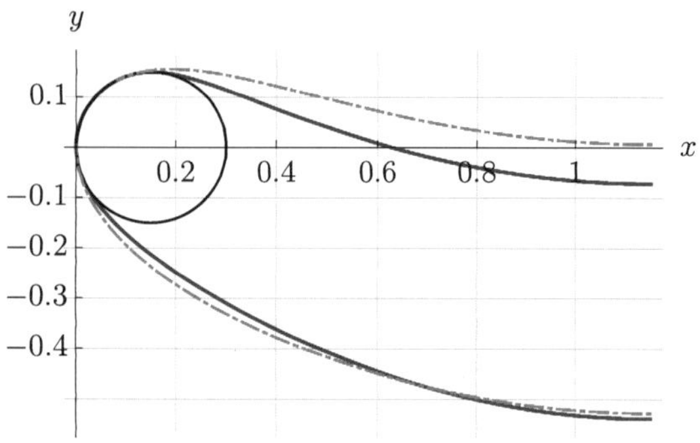

Figure 4. Belt and pulley. *Solid line* is the belt current configuration, $\rho = 0.48\,\mathrm{kg/m}$; *dot-dashed line* for $\rho = 0.16\,\mathrm{kg/m}$, *thin circle* is the pulley.

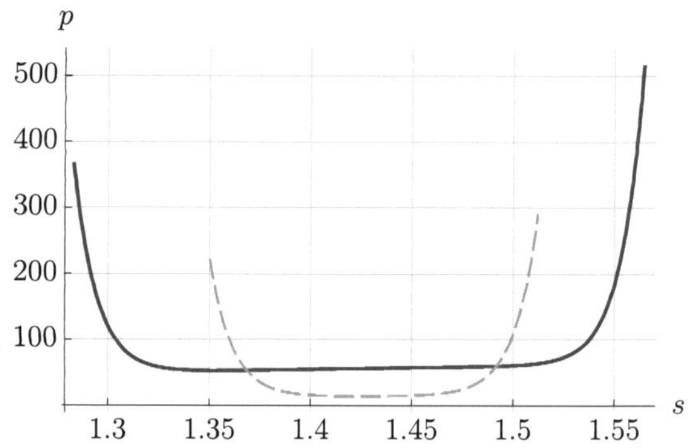

Figure 5. Pressure distribution vs material coordinate s. *Solid line* for $\rho = 0.48\,\mathrm{kg/m}$, *dashed line* for $\rho = 0.16\,\mathrm{kg/m}$.

The pressure distribution is symmetric in the case without gravity, and it tends to concentrate at one (upper) side with gravity included. In figure 6, 7 we show the forces P_1, P_2 acting at the middle of the spans and the contact angles β_1, β_2 (central angles of contact zones on pulleys, see figure 2). For $y^{(1)}\big|_{\xi=1} < 0$ we have $\beta_1 > 0$, and for $y^{(3)}\big|_{\xi=1} > 0$ we have $\beta_2 > 0$. The lower span has lower tension force, than the upper one. The gravity force may significantly influence the angle of contact $\beta_2 + \beta_1$ and, therefore, the friction behaviour of the belt drive.

7. Conclusion

The main results of the present study are listed below:

- a system of ODE for the drive belt as a plane nonlinear elastic rod with account for tension and transverse shear is derived;
- the considered part of the belt is divided into three segments, the contact segment and two free segments;
- the equations for the free and contact segments which have qualitatively different formulation are combined in a single system of ODE;
- the formulated BVP is suitable for numerical solving by the shooting, finite difference or collocation method;
- the effect of gravity force acting on the belt is clarified.

The generalization to the cases with pulleys of unequal radii and an inclination of the gravity force relative to y-axis is possible. These results will be used in further work including the

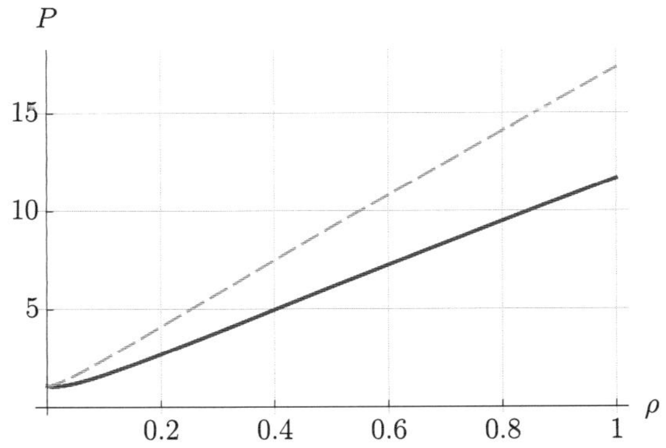

Figure 6. Forces at the middle of spans vs belt mass density. *Solid line* is the force of lower span P_1, *dashed line* is the force of upper span P_2.

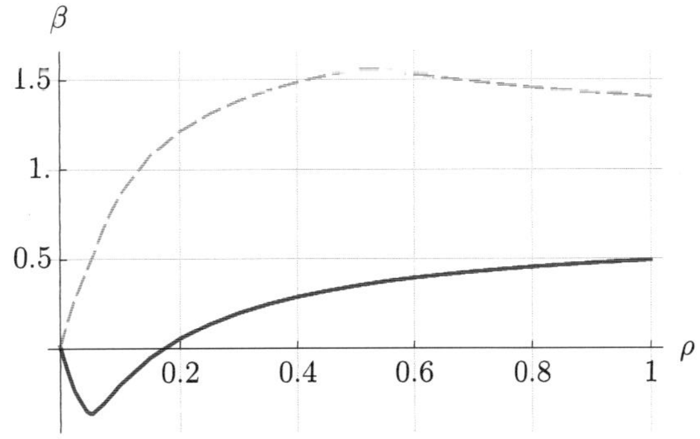

Figure 7. Contact angles vs belt mass density. *Solid line* is the angle of lower span β_1, *dashed line* is the angle of upper span β_2.

study of belt-pulley interaction with friction and the dynamical modeling, both stationary and transient.

Acknowledgments
This research is carried out in the framework of the joint project of the Russian Foundation for Basic Research (Grant No. 14–51–15001) and the Austrian Science Fund (Grant No. I 2093 International Project).

References

[1] Reynolds O 1874 *The Engineer* **38** 396
[2] Eliseev V and Vetyukov Y 2012 *Acta Mechanica* **223** 1657–1667
[3] Rubin M B 2000 *J Mech. Des.* **122** 311–316
[4] Vetyukov Y, Oborin E, Krommer M and Eliseev V 2017 *Mathematical and Computer Modelling of Dynamical Systems* **23** 40–54
[5] Belyaev A, Eliseev V, Irschik H and Oborin E 2016 *PAMM* **16** 11–14
[6] Belyaev A, Eliseev V, Irschik H and Oborin E 2017 *Dynamics & Control of Advanced Structures and Machines* ed Irschik H, Belyaev A and Krommer M (Springer International Publishing) pp 195–203
[7] Gerbert G 1999 *Traction Belt Mechanics* (Göteborg: Chalmers University of Technology)
[8] Kong L and Parker R G 2005 *J. Appl. Mech.* **72** 25–34
[9] Morimoto T and Iizuka H 2012 *Int. J. Mech. Sci.* **54** 234–240
[10] Pechstein A and Gerstmayr J 2013 *Multibody System Dynamics* **30** 343–358
[11] Antman S S and Schuricht F 2003 *Math. & Mech. of Solids* **8** 75–102
[12] Chen J S 2011 *Acta Mechanica* **222** 225–232
[13] Eliseev V and Zinovieva T 2012 *Comp. Cont. Mech.* **5** 70–78
[14] Gasmi A, Joseph P F, Rhyne T B and Cron S M 2012 *Int. J. Solids & Structures* **49** 2604–2616
[15] Osipenko M, Nyashin Y I and Kasatkin A 2015 *PRNPU Mechanics Bulletin* 121–129
[16] Berger E 2002 *Applied Mech. Reviews* **55** 535–577
[17] Menq C H, Bielak J and Griffin J 1986 *J. Sound & Vibration* **107** 279–293
[18] Alciatore D and Traver A 1995 *ASME J. Mech. Des.* **117** 506–511
[19] Belyaev A K, Eliseev V V and Oborin E A 2016 *Int. Review Mech. Eng.* **10** 333–338
[20] Firbank T 1970 *Int. J. Mech. Sci.* **12** 1053–1063
[21] Antman S 2005 *Nonlinear Problems of Elasticity* (New York: Springer)
[22] Eliseev V 1988 *J Appl. Math. & Mech.* **52** 493–498
[23] Eliseev V 2006 *Mechanics of Deformable Solid Bodies* (St. Petersburg: St.-Peterbg. State Polytech. Univ. Publ. House)
[24] Wriggers P and Laursen T A 2006 *Computational Contact Mechanics* (Berlin: Springer)

[25] Durville D 2012 *Comp. Mech.* **49** 687–707
[26] Denoël V and Detournay E 2016 *Int. J. Solids & Structures* **48** 625–636
[27] Huynen A, Detournay E and Denoël V 2016 *Proc. R. Soc. A* **472** 20150547
[28] Liakou A and Detournay E to appear in 2018 *Int. J. Non-Linear Mech.*
[29] Belyaev A, Eliseev V, Irschik H and Oborin E 2017 *Acta Mechanica* **228** 4425–4434
[30] Belyaev A K, Eliseev V V, Irschik H and Oborin E 2017 *IOP Conf. Series: Mater. Sci. Eng.* **208** 012004
[31] Vetyukov Y 2017 *Analysis and Modelling of Advanced Structures and Smart Systems* (*Advanced Structured Materials* vol 81) ed Altenbach H, Carrera E and Kulikov G (Singapore: Springer) pp 291–325
[32] Vetyukov Y 2018 *J. Sound & Vibration* **414** 299–317
[33] Oborin E, Vetyukov Y and Steinbrecher I 2018 *International Journal of Solids and Structures*
[34] Ascher U and Russell R D 1981 *SIAM Review* **23** 238–254
[35] Kaczmarczyk S and Mirhadizadeh S 2016 *J. Phys.: Conf. Series* **721** 012011
[36] Driscoll T, Hale N and Trefethen L (eds) 2014 *Chebfun Guide* (Oxford: Pafnuty Publications)

Comparison of Three Different Techniques for Nonlinear Position and Vibration Control of a Rotating Smart Beam

A Fenili

Federal University of ABC, Aerospace Engineering / Mechanical Engineering,
Santo André, SP, Brazil.

Abstract. In this paper, position and vibration control of a smart rotating flexible beam-like structure is investigated. The nonlinear integro-partial differential governing equations of motion are derived via the Extended Hamilton's Principle. A State-Dependent Riccati Equation (SDRE) based methodology is used in order to design a nonlinear controller. Three different techniques are considered for comparison: (1) control torque is applied at the rotating axis for angular position and vibration control (2) control torque is applied at the rotating axis for angular position control and a distributed moment is applied via piezoelectric actuator for vibration control and (3) control torque is applied at the rotating axis for angular position and vibration control and a distributed moment is applied via piezoelectric actuator for vibration control.

1. Introduction

Flexible beam-like structures usually demand low energy consumption, deliver increased performance, and are easily deployed and assembled. Notwithstanding these benefits, such structures are more difficult to control and stabilize because of their susceptibility to vibration. A variation of mathematical models can be investigated with different degrees of complexity in order to represent a rotating flexible beam. In [1], for example, the authors derive a nonlinear coupled set of PDEs that governs a rotating composite Timoshenko beam with a preset pitch angle.

Investigations on the use of piezoelectric actuators to control vibration of flexible structures are vast [2, 3]. The most effective position to place these actuators in a clamped-free beam is at the root of the structure [2]. The piezoelectric ceramic is considered a smart material because its properties can be changed in a controlled fashion by external stimuli [3]. The use of the piezoelectric elements adhered to the surface of the structure does not add significant weight to the original structure.

State-Dependent Riccati Equation (SDRE) is a strategy for nonlinear optimal control design that synthesizes nonlinear feedback controllers by solving an infinite-horizon nonlinear optimal regulation problem for a system full-state observable, autonomous, nonlinear in the state, and affine in the input [4].

2. Governing equations of motion: the rotating flexible beam plus piezoelectric actuator

The geometric model of the system investigated in this work is presented in figure 1. This system comprises a flexible beam-like structure connected to a rotating rigid hub.

The governing equations of motion are obtained via the Extended Hamilton's principle [5]. The kinetic energy, T, of the system is given by:

$$\mathrm{T} = \frac{1}{2}\mathrm{I_h}\dot{\theta}^2 + \frac{1}{2}\int_0^{\ell} \rho \mathrm{A}\,[\,(\dot{v} + x\dot{\theta})^2 + (v\dot{\theta})^2\,]\mathrm{d}x \tag{1}$$

IOP Conf. Series: Journal of Physics: Conf. Series **1048** (2018) 012003 doi:10.1088/1742-6596/1048/1/012003

In equation (1), I_h represents the moment of inertia of the hub about the rotational axis, $\theta(t)$ represents the angular displacement of the system, $v(x,t)$ represents the transversal displacement of the beam, ρ represents the mass density of the beam, A represents the cross-section area of the beam and ℓ represents the length of the undeflected beam.

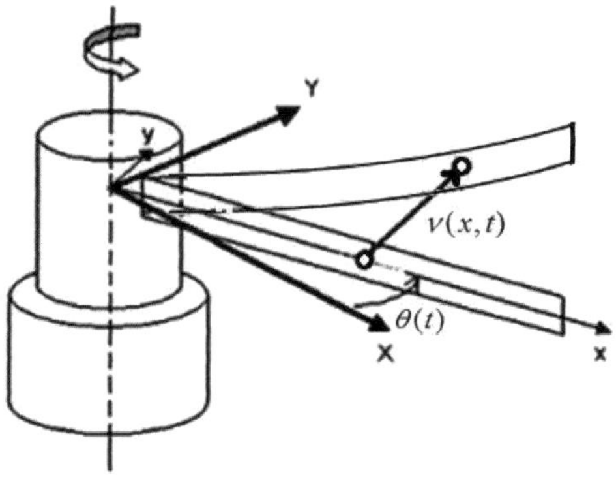

Figure 1. The slewing flexible beam system (XY: inertial axis; xy: moving axis).

Linear curvature is assumed in the mathematical modeling of the flexible structure [5]. The potential (strain) energy, V, of the system is given by:

$$V = \frac{1}{2}\int_0^\ell EI v''^2 dx \tag{2}$$

In equation (2), E represents the Young's modulus of the material of the beam, and I represents the moment of inertia of the cross-section area of the beam. The work done by the external torque and the piezoelectric forces is given by:

$$W = T_{rot}\theta + \int_0^\ell f(x,t)v(x,t)\,dx \tag{3}$$

In equation (3), T_{rot} represents a external torque applied to the rotating axis and f(x,t) is given by:

$$f(x,t) = \frac{\partial^2 M_{piezo}}{\partial x^2} \tag{4}$$

In equation (4) [3, 6],

$$M_{piezo} = \frac{1}{2}E_c d_{31} b(t_b + t_c)V_a(x,t) \tag{5}$$

In equation (5), $V_a(x, t)$ is the control voltage applied to the actuator, E_c is the elastic modulus of the piezoceramic, d_{31} is a piezoelectric constant, b is the piezoceramic length, t_b is the beam thickness and t_c is the piezoceramic thickness.
The Extended Hamilton's principle states that:

$$\delta\int_{t_1}^{t_2}(T - V + W)\,dt = 0 \tag{6}$$

In equation (6), $\delta(t_1) = \delta(t_2) = 0$.

Substituting equations (1), (2) and (3) in equation (6) (and making some manipulations) one obtains the nonlinear integro-partial differential set of equations of the smart structure (flexible structure + piezoelectric actuator) given by:

$$\ddot{\theta}+2\dot{\theta}\alpha_1+\alpha_2\dot{\theta}^2-\frac{\mathrm{EI}}{\rho\mathrm{A}}\alpha_3=\alpha_4\mathrm{T}_{\mathrm{rot}}-\frac{\mathrm{C_a}}{\rho\mathrm{A}\alpha_0}\int_0^{\mathrm{L}} xV_a''\mathrm{d}x \tag{7}$$

$$\ddot{v}+\frac{\mathrm{EI}}{\rho\mathrm{A}}\left(v^{iv}+x\alpha_3\right)-2x\dot{\theta}\alpha_1-\left(v+x\alpha_2\right)\dot{\theta}^2=-x\alpha_4\mathrm{T}_{\mathrm{rot}}+\frac{\mathrm{C_a}}{\rho\mathrm{A}}\left(V_a''+\frac{x}{\alpha_0}\int_0^{\mathrm{L}} xV_a''\mathrm{d}x\right) \tag{8}$$

and the boundary condition for the clamped-free as beam given by:

$$v\,(0,t)=0 \tag{9}$$

$$v'\,(0,t)=0 \tag{10}$$

$$v''\,(\mathrm{L},t)=0 \tag{11}$$

$$v'''\,(\mathrm{L},t)=0 \tag{12}$$

In equations (7) and (8), the parameters α_i are given by:

$$\alpha_0=\frac{\mathrm{I_{hub}}}{\rho\mathrm{A}}+\int_0^{\mathrm{L}} v^2\mathrm{d}x \tag{13}$$

$$\alpha_1=\frac{1}{\alpha_0}\int_0^{\mathrm{L}} v\dot{v}\,\mathrm{d}x \tag{14}$$

$$\alpha_2=\frac{1}{\alpha_0}\int_0^{\mathrm{L}} vx\,\mathrm{d}x \tag{15}$$

$$\alpha_3=\frac{1}{\alpha_0}\int_0^{\mathrm{L}} xv^{iv}\mathrm{d}x \tag{16}$$

$$\alpha_4=\frac{1}{\rho\mathrm{A}\alpha_0} \tag{17}$$

The time-varying parameters given by equations (13) to (17) contain definite integral terms. In order to proceed with the numerical integration of equations (7) and (8), these terms are expanded using the method for numerical approximation of definite integrals named Simpson's rule [7] and given by:

$$\int_a^b f(x)\mathrm{d}x\approx\frac{h}{3}\left(f(x_0)+4f(x_1)+2f(x_2)+4f(x_3)+2f(x_4)+\ldots+4f(x_{n-1})+f(x_n)\right) \tag{18}$$

Consider that the beam is divided in 8 elements (or 9 nodes). The element size is h. Using this approximation rule and this number of elements, equations (13) to (16) – equation (17) is not necessary to discuss because it uses equation (13) – can be rewriten as:

$$\alpha_0=\frac{\mathrm{I_{hub}}}{\rho\mathrm{A}}+\frac{h}{3}\left(v_1^2+4v_2^2+2v_3^2+4v_4^2+2v_5^2+4v_6^2+2v_7^2+4v_8^2+v_9^2\right) \tag{19}$$

$$\alpha_1 = \frac{1}{\alpha_0}\frac{h}{3}\left(v_1\dot{v}_1 + 4v_2\dot{v}_2 + 2v_3\dot{v}_3 + 4v_4\dot{v}_4 + 2v_5\dot{v}_5 + 4v_6\dot{v}_6 + 2v_7\dot{v}_7 + 4v_8\dot{v}_8 + v_9\dot{v}_9 \right) \tag{20}$$

$$\alpha_2 = \frac{1}{\alpha_0}\frac{h}{3}\left(4hv_2 + 4hv_3 + 12hv_4 + 8hv_5 + 20hv_6 + 12hv_7 + 28hv_8 + Lv_9 \right) \tag{21}$$

$$\alpha_3 = \frac{1}{\alpha_0}\frac{h}{3}\left(4h\frac{\partial^4 v_2}{\partial x^4} + 4h\frac{\partial^4 v_3}{\partial x^4} + 12h\frac{\partial^4 v_4}{\partial x^4} + 8h\frac{\partial^4 v_5}{\partial x^4} + 20h\frac{\partial^4 v_6}{\partial x^4} + 12h\frac{\partial^4 v_7}{\partial x^4} + 28h\frac{\partial^4 v_8}{\partial x^4} + L\frac{\partial^4 v_9}{\partial x^4} \right) \tag{22}$$

After the substitution of the numerical approximation presented in equations (19) to (22), the resulting nonlinear partial-differential governing equations of motion are transformed into a set of nonlinear ordinary differential equations via a finite difference approach [8]. The division of the beam in elements is given according to figure 2.

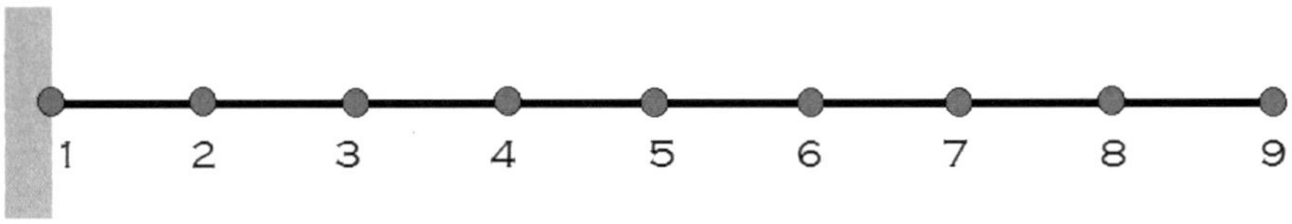

Figure 2. Division of the beam in elements for the application of the finite difference approach.

Considering nodes 2 to 9, the central difference formula using 5 points is considered. This formula for the fourth derivative is given by:

$$\frac{\partial^4 v_i(x_i,t)}{\partial x^4} = \frac{v(x_i+2h,t) - 4v(x_i+h,t) + 6v(x_i,t) - 4v(x_i-h,t) + v(x_i-2h,t)}{h^4} \tag{23}$$

For nodes 2 and 9, the boundary conditions are used to eliminate the points located outside the beam. The fourth order partial derivatives in equation (22) are also approximated using equation (23). Using the Simpson´s rule, the integral in the right sides of equations (7) and (8) can be writen as:

$$\int_0^L xV_a''\mathrm{d}x = \frac{h}{3}\left(x_1\frac{\partial^2 V_{a1}}{\partial x^2} + 4x_2\frac{\partial^2 V_{a2}}{\partial x^2} + 2x_3\frac{\partial^2 V_{a3}}{\partial x^2} + 4x_4\frac{\partial^2 V_{a4}}{\partial x^2} + 2x_5\frac{\partial^2 V_{a5}}{\partial x^2} + 4x_6\frac{\partial^2 V_{a6}}{\partial x^2} + 2x_7\frac{\partial^2 V_{a7}}{\partial x^2} + 4x_8\frac{\partial^2 V_{a8}}{\partial x^2} + x_9\frac{\partial^2 V_{a9}}{\partial x^2} \right) \tag{24}$$

The finite difference approach considering the central difference formula using 3 points is considered in order to rewrite the second derivatives in equation (24). This formula is given by:

$$\frac{\partial^2 V_{an}(x,t)}{\partial x^2} = \frac{V_{an-1} - 2V_{an} + V_{an+1}}{h^2} \tag{25}$$

The final expression, then, for equation (24) is:

$$\int_0^L xV_a''\mathrm{d}x = \frac{4}{3}V_{a1} - \frac{4}{3}V_{a2} + \frac{8}{3}V_{a3} - 4V_{a4} + \frac{16}{3}V_{a5} - \frac{20}{3}V_{a6} + 8V_{a7} - 12V_{a8} + 4V_{a9} \tag{26}$$

In this work only two piezoelements (regarding three nodes) are considered, according to figure 3.

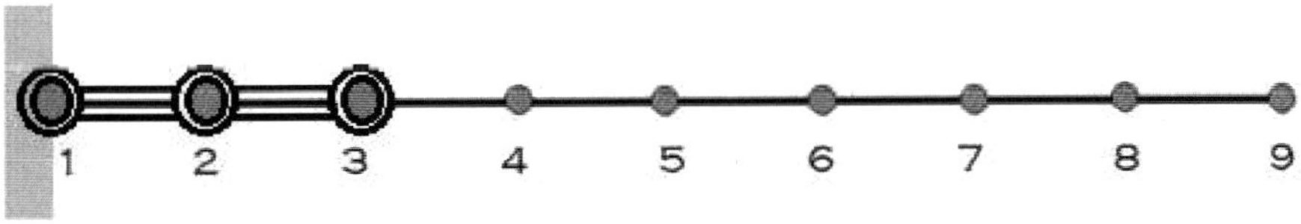

Figure 3. Division of the beam in elements with the location of the piezoelectric actuator (nodes 1 to 3).

Considering only two elements near the beam root, equation (26) van be reduced to:

$$\int_0^L x V_a'' \, dx = \frac{4}{3}V_{a1} - \frac{4}{3}V_{a2} + \frac{8}{3}V_{a3} \tag{27}$$

After all these manipulations, the complete set of nonlinear ordinary differential governing equations of motion for the rotating flexible structure with piezoelectric elements attached is finally given by:

$$\ddot{\theta} + \beta_1\dot{\theta} + \beta_2\dot{\theta}^2 - \beta_{3a}v_2 - \beta_{3b}v_3 - \beta_{3c}v_4 - \beta_{3d}v_5 - \beta_{3e}v_6 - \beta_{3f}v_7 - \beta_{3g}v_8 - \beta_{3h}v_9 = \beta_4 \mathrm{T}_{\mathrm{rot}} - \beta_5\left(V_{a1} - V_{a2} + 2V_{a3}\right) \tag{28}$$

$$\ddot{v}_2 + (7\beta_0 + \beta_{3a}x_2)v_2 - (4\beta_0 - \beta_{3b}x_2)v_3 + (\beta_0 + \beta_{3c}x_2)v_4 + \beta_{3d}x_2v_5 + \beta_{3e}x_2v_6 + \beta_{3f}x_2v_7 + \beta_{3g}x_2v_8 + \beta_{3h}x_2v_9 - \beta_1x_2\dot{\theta} - (v_2 + \beta_2x_2)\dot{\theta}^2 = -\beta_4x_2\mathrm{T}_{\mathrm{rot}} + \left(\frac{\beta_6}{h^2} + \beta_5x_2\right)V_{a1} - \left(\frac{2\beta_6}{h^2} + \beta_5x_2\right)V_{a2} + \left(\frac{\beta_6}{h^2} + 2\beta_5x_2\right)V_{a3} \tag{29}$$

$$\ddot{v}_3 - (4\beta_0 - \beta_{3a}x_3)v_2 + (6\beta_0 + \beta_{3b}x_3)v_3 - (4\beta_0 - \beta_{3c}x_3)v_4 + (\beta_0 + \beta_{3d}x_3)v_5 + \beta_{3e}x_3v_6 + \beta_{3f}x_3v_7 + \beta_{3g}x_3v_8 + \beta_{3h}x_3v_9 - \beta_1x_3\dot{\theta} - (v_3 + \beta_2x_3)\dot{\theta}^2 = -\beta_4x_3\mathrm{T}_{\mathrm{rot}} + (\beta_5x_3)V_{a1} + \left(\frac{\beta_6}{h^2} - \beta_5x_3\right)V_{a2} - \left(\frac{2\beta_6}{h^2} - 2\beta_5x_3\right)V_{a3} \tag{30}$$

$$\ddot{v}_4 + (\beta_0 + \beta_{3a}x_4)v_2 - (4\beta_0 - \beta_{3b}x_4)v_3 + (6\beta_0 + \beta_{3c}x_4)v_4 - (4\beta_0 - \beta_{3d}x_4)v_5 + (\beta_0 + \beta_{3e}x_4)v_6 + \beta_{3f}x_4v_7 + \beta_{3g}x_4v_8 + \beta_{3h}x_4v_9 - \beta_1x_4\dot{\theta} - (v_4 + \beta_2x_4)\dot{\theta}^2 = -\beta_4x_4\mathrm{T}_{\mathrm{rot}} \tag{31}$$

$$\ddot{v}_5 + \beta_{3a}x_5v_2 + (\beta_0 + \beta_{3b}x_5)v_3 - (4\beta_0 - \beta_{3c}x_5)v_4 + (6\beta_0 + \beta_{3d}x_5)v_5 - (4\beta_0 - \beta_{3e}x_5)v_6 + (\beta_0 + \beta_{3f}x_5)v_7 + \beta_{3g}x_5v_8 + \beta_{3h}x_5v_9 - \beta_1x_5\dot{\theta} - (v_5 + \beta_2x)_5\dot{\theta}^2 = -\beta_4x_5\mathrm{T}_{\mathrm{rot}} \tag{32}$$

$$\ddot{v}_6 + \beta_{3a}x_6v_2 + \beta_{3b}x_6v_3 + (\beta_0 + \beta_{3c}x_6)v_4 - (4\beta_0 - \beta_{3d}x_6)v_5 + (6\beta_0 + \beta_{3e}x_6)v_6 - (4\beta_0 - \beta_{3f}x_6)v_7 + (\beta_0 + \beta_{3g}x_6)v_8 + \beta_{3h}x_6v_9 - \beta_1x_6\dot{\theta} - (v_6 + \beta_2x_6)\dot{\theta}^2 = -\beta_4x_6\mathrm{T}_{\mathrm{rot}} \tag{33}$$

$$\ddot{v}_7 + \beta_{3a}x_7v_2 + \beta_{3b}x_7v_3 + \beta_{3c}x_7v_4 + (\beta_0 + \beta_{3d}x_7)v_5 - (4\beta_0 - \beta_{3e}x_7)v_6 + (6\beta_0 + \beta_{3f}x_7)v_7 - (4\beta_0 - \beta_{3g}x_7)v_8 + (\beta_0 + \beta_{3h}x_7)v_9 - \beta_1x_7\dot{\theta} - (v_7 + \beta_2x_7)\dot{\theta}^2 = -\beta_4x_7\mathrm{T}_{\mathrm{rot}} \tag{34}$$

$$\ddot{v}_8 + \beta_{3a}x_8v_2 + \beta_{3b}x_8v_3 + \beta_{3c}x_8v_4 + \beta_{3d}x_8v_5 + (\beta_0 + \beta_{3e}x_8)v_6 - (4\beta_0 - \beta_{3f}x_8)v_7 + (5\beta_0 + \beta_{3g}x_8)v_8 - (2\beta_0 - \beta_{3h}x_8)v_9 - \beta_1x_8\dot{\theta} - (v_8 + \beta_2x_8)\dot{\theta}^2 = -\beta_4x_8\mathrm{T}_{\mathrm{rot}} \tag{35}$$

IOP Conf. Series: Journal of Physics: Conf. Series **1048** (2018) 012003 doi:10.1088/1742-6596/1048/1/012003

$$\ddot{v}_9 + \beta_{3a}x_9v_2 + \beta_{3b}x_9v_3 + \beta_{3c}x_9v_4 + \beta_{3d}x_9v_5 + \beta_{3e}x_9v_6 + (2\beta_0 + \beta_{3f}x_9)v_7 - (4\beta_0 - \beta_{3g}x)_9 v_8 + (2\beta_0 + \beta_{3h}x_9)v_9 - \beta_1 x_9\dot{\theta} - (v_9 + \beta_2 x_9)\dot{\theta}^2 = -\beta_4 x_9 \mathrm{T}_{\mathrm{rot}} \tag{36}$$

In equations (28) to (36):

$$\gamma = \frac{\mathrm{I}_{\mathrm{hub}}}{\rho\mathrm{A}} + \frac{h}{3}\left(4v_2^2 + 2v_3^2 + 4v_4^2 + 2v_5^2 + 4v_6^2 + 2v_7^2 + 4v_8^2 + v_9^2\right) \qquad \beta_0 = \frac{\mathrm{EI}}{\rho\mathrm{A}h^4} \qquad \eta = \frac{\mathrm{EI}}{3\rho\mathrm{A}h^2}\left(\frac{1}{\gamma}\right)$$

$$\beta_1 = \frac{2h}{3}\left(\frac{4v_2\dot{v}_2 + 2v_3\dot{v}_3 + 4v_4\dot{v}_4 + 2v_5\dot{v}_5 + 4v_6\dot{v}_6 + 2v_7\dot{v}_7 + 4v_8\dot{v}_8 + v_9\dot{v}_9}{\gamma}\right)$$

$$\beta_2 = \frac{4h^2}{3}\left(\frac{v_2 + v_3 + 3v_4 + 2v_5 + 5v_6 + 3v_7 + 7v_8 + \dfrac{L}{4h}v_9}{\gamma}\right) \qquad \beta_{3a} = 24\eta \qquad \beta_{3b} = -32\eta$$

$$\beta_{3c} = 48\eta \qquad \beta_{3d} = -64\eta \qquad \beta_{3e} = 80\eta \qquad \beta_{3f} = \eta\left(\frac{2\mathrm{L}}{h} - 112\right) \qquad \beta_{3g} = \eta\left(112 - \frac{4\mathrm{L}}{h}\right)$$

$$\beta_{3h} = \eta\left(\frac{2\mathrm{L}}{h} - 44\right) \qquad \beta_4 = \frac{1}{\rho\mathrm{A}}\left(\frac{1}{\gamma}\right) \qquad \beta_5 = \frac{4C_a}{3\rho\mathrm{A}\alpha_0} \qquad \beta_6 = \frac{C_a}{\rho\mathrm{A}}$$

This set of governing equations is now up to be put in state space form and numerically integrated using the numerical integration method named fourth order Runge-Kutta.

3. The State-Dependent Riccati Equation (SDRE) control

The SDRE (State-Dependent Riccati Equation) approach to nonlinear system control relies on representing a nonlinear system's dynamics with state-dependent matrices and therefore state-dependent Riccati equations in order to generate a feedback law [4]. The nonlinear system:

$$\dot{\boldsymbol{x}} = \boldsymbol{f}(\boldsymbol{x}) + \boldsymbol{B}(\boldsymbol{x})\boldsymbol{u} \tag{37}$$

can be rewriten in the form:

$$\dot{\boldsymbol{x}} = \boldsymbol{A}(\boldsymbol{x})\boldsymbol{x} + \boldsymbol{B}(\boldsymbol{x})\boldsymbol{u} \tag{38}$$

The feedback law is given by [4]:

$$\boldsymbol{u} = -\boldsymbol{R}^{-1}(\boldsymbol{x})\boldsymbol{B}^{\mathrm{T}}(\boldsymbol{x})\boldsymbol{P}(\boldsymbol{x})\boldsymbol{x} \tag{39}$$

and minimize the infinite-time performance criterion:

$$J = \frac{1}{2}\int_0^\infty \left\{\boldsymbol{x}^{\mathrm{T}}\boldsymbol{Q}(\boldsymbol{x})\boldsymbol{x} + \boldsymbol{u}^{\mathrm{T}}\boldsymbol{R}(\boldsymbol{x})\boldsymbol{u}\right\}\mathrm{d}t \tag{40}$$

subject to the input-affine nonlinear differential constraint given by equation (37) (or equation (38)) while regulating the system to the origin such that $\lim_{t\to\infty}\boldsymbol{x}(t) = 0$. This methodology forms the basis of the SDRE method for nonlinear regulation.

In equation (39), $\boldsymbol{P}(\boldsymbol{x})$ is obtained from the solution of the state dependent Riccati equation:

$$\boldsymbol{P}(\boldsymbol{x})\boldsymbol{A}(\boldsymbol{x}) + \boldsymbol{A}^{\mathrm{T}}(\boldsymbol{x})\boldsymbol{P}(\boldsymbol{x}) + \boldsymbol{Q}(\boldsymbol{x}) - \boldsymbol{P}(\boldsymbol{x})\boldsymbol{B}(\boldsymbol{x})\boldsymbol{R}^{-1}(\boldsymbol{x})\boldsymbol{B}^{\mathrm{T}}(\boldsymbol{x})\boldsymbol{P}(\boldsymbol{x}) = \boldsymbol{0} \tag{41}$$

In equation (41), Q(x) and R(x) are design parameters that satisfy the positive definiteness conditions $\boldsymbol{Q}(\boldsymbol{x}) > \boldsymbol{0}$ and $\boldsymbol{R}(\boldsymbol{x}) > \boldsymbol{0}$.

In this work, the control vector, $\boldsymbol{u}$, comprises T_{rot} and the voltages V_{a1}, V_{a2} and V_{a3} applied to the first three nodes of the beam.

4. Numerical simulations

The parameters used in the numerical simulations are presented in table 1. The piezoactuator width is the same as the beam width. The values chosen for the piezoactuator represents a real piezoactuator. The values chosen for the beam represents a feasible aluminum flexible beam.

Table 1. Parameters used in the numerical simulations.

Parameter	Nomenclature	Value
Mass density (beam)	ρ	2700 kg/m^3
Elastic modulus (beam)	E	0.700*10^{11} Pa
Beam length	L	1.500 m
Beam thickness	h_b	1.000*10^{-3} m
Beam width	w	2.000*10^{-2} m
Elastic modulus (piezoactuator)	E_c	0.714*10^{11} Pa
Piezoactuator strain constant	d_{31}	200*10^{-12} m/V
Piezoactuator thickness	t_c	0.305*10^{-3} m
Piezoactuator length	b	0.1875 m

The weight matrices Q and R used in this work are composed by constant elements and given by:

$$R=\begin{bmatrix}1 & 0 & 0 & 0\\ 0 & 0.01 & 0 & 0\\ 0 & 0 & 0.01 & 0\\ 0 & 0 & 0 & 0.01\end{bmatrix} \qquad Q=\begin{bmatrix}70 & 0 & 0 & 0 & \cdots & 0\\ 0 & 70 & 0 & 0 & \cdots & 0\\ 0 & 0 & 140 & 0 & \cdots & 0\\ 0 & 0 & 0 & 140 & \cdots & 0\\ \vdots & \vdots & \vdots & \vdots & \ddots & \vdots\\ 0 & 0 & 0 & 0 & 0 & 140\end{bmatrix}_{18X18}$$

The time step used in the numerical integration of the governing equations via the fourth order Runge-Kutta is 0.001s.

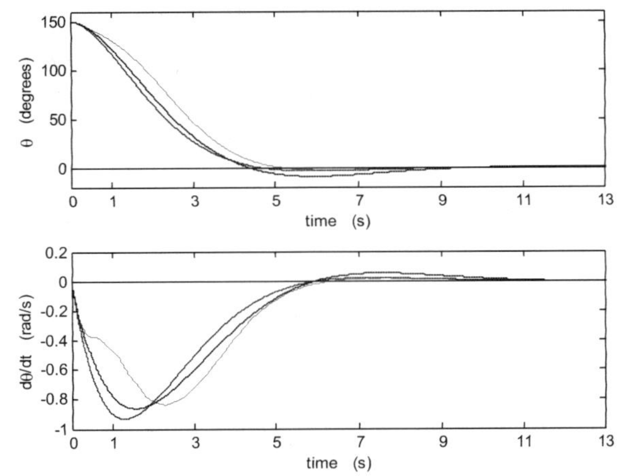

Figure 4. Angular displacement and velocity: case 1 (——), case 2 (·-·-·) and case 3 (- - - -).

Three cases are investigated here. In case 1, a control torque is applied at the rotating axis for angular position and vibration control. In case 2, a control torque is applied at the rotating axis for angular position control and a moment is applied along the flexible structure via the piezoelectric actuator for vibration control. In case 3, a control torque is applied at the rotating axis for angular position and vibration control and a moment is also applied along the flexible structure via the piezoelectric actuator for vibration control.

According to figure 4, there is no meaningfully difference in the amplitudes of the angular displacements for the three cases.

The differences in the amplitudes of the angular velocities are more remarkable. The maximum values of the amplitudes of the angular velocity for cases 1 and 3 are basically the same (although the velocity profiles for these cases are very different) and both values are smaller when compared to the maximum value of the angular velocity for case 2.

The case 2 is the only case where the control torque applied to the rotating axis is not used for the flexible structure vibration control.

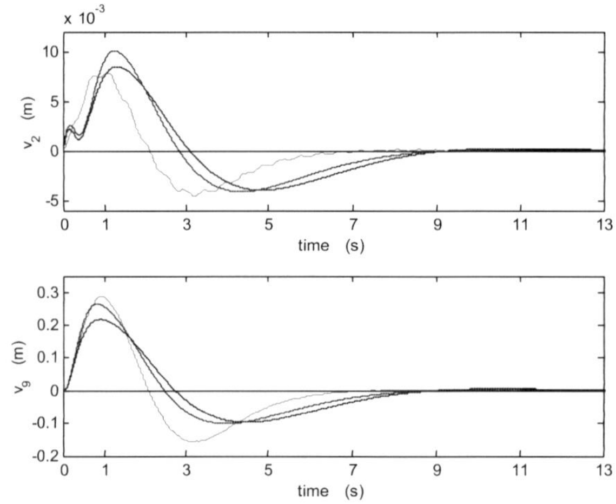

Figure 5. Vibration control in nodes 2 and 9: case 1 (——), case 2 (-·-·-) and case 3 (- - - -).

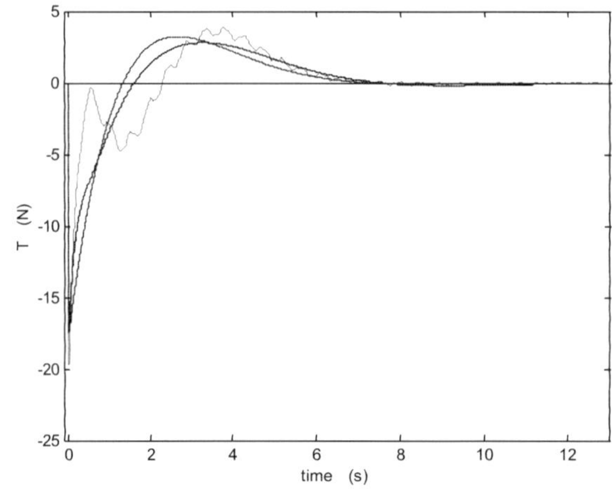

Figure 6. Control torque applied to the rotating axis: case 1 (——), case 2 (-·-·-) and case 3 (- - - -).

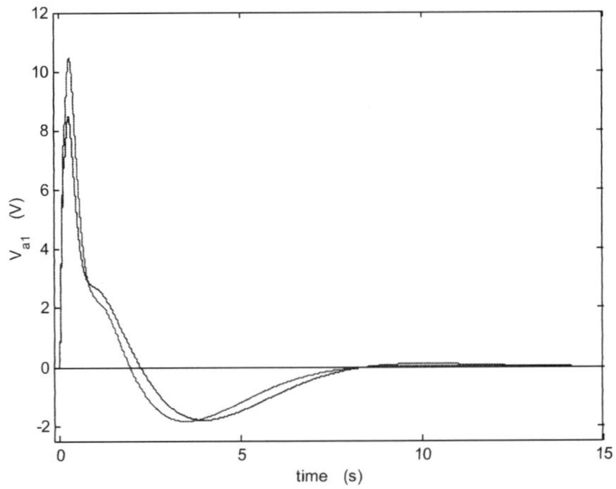

Figure 7. Control voltage on the piezoelement: case 2 (¨·¨·¨) and case 3 (- - - -) – first node.

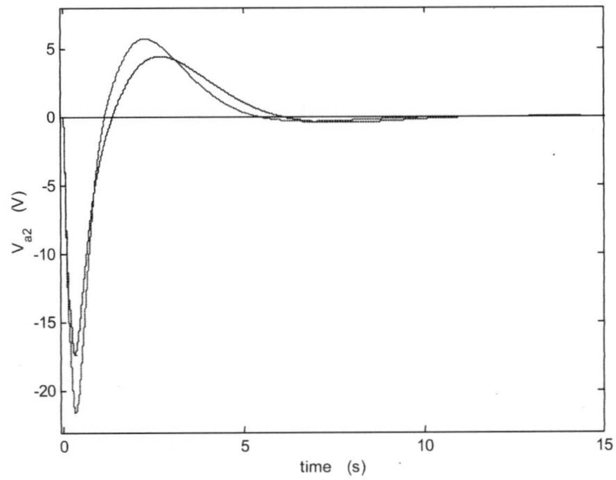

Figure 8. Control voltage on the piezoelement: case 2 (¨·¨·¨) and case 3 (- - - -) – second node.

According to figure 5, the maximum amplitude of the response of node 9 for case 3 is about 7.5 centimeters smaller than the maximum amplitude of the response of the same node for case 1 and about 5 centimeters smaller than the maximum amplitude of the response for case 2. In this sense, for vibration control, case 3 presents the better results.

For node 2, the differences in the amplitudes of the response between the three cases are not significant.

Other reasons why case 3 can be considered the better solution can be verified in figures 6 to 9.

In figure 6, the maximum amplitudes for the control torque applied at the rotating axis has smaller values for case 3.

In figures 7, 8 and 9 it is clear that for nodes 1 to 3 (the three nodes related to the location of the piezoelement along the beam) the control voltage for case 3 presents the smallest values when compared to case 2.

The last two paragraphs means that in case 3 one needs less energy to drive the system to the desired final states.

IOP Conf. Series: Journal of Physics: Conf. Series **1048** (2018) 012003 doi:10.1088/1742-6596/1048/1/012003

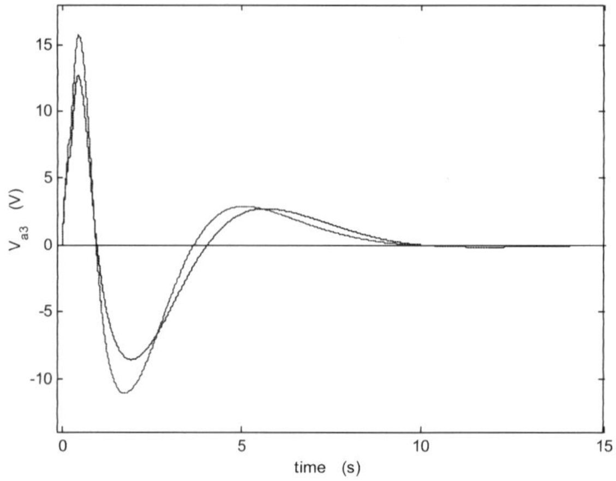

Figure 9. Control voltage on the piezoelement: case 2 (`-.-.-`) and case 3 (- - - -) – third node.

5. Conclusions

The numerical solutions presented in this work show that the SDRE control is effective when used for position and vibration control in a rotating flexible beam-like system actuated by external torque and piezoelectric moments. Three cases are studied here based on control distribution. For position control all the three cases present very similar results. The most effective case for vibration control is the one where the beam vibration is controlled by both the piezoelement and the external torque applied to the rotating axis. In the cases studied here, only 25% of beam length is covered by the piezoelement. The results presented here can be improved by adding more piezoelements to the beam.

6. References

[1] Georgiades F, Latalski J and Warminski J 2014 Equations of motion of rotating composite beams with a no constant rotation speed and an arbitrary preset angle *Meccanica* **49** issue 8 1833–58.

[2] Gani A, Salami-SMIEEE M J E and Khan Md R 2003 Active vibration control of a beam with piezoelectric patches: real-time implementation with xPC target *Proceedings of 2003 IEEE Conference on Control Applications - CCA, Istanbul* 538–44.

[3] Zhang J, He L and Wang E 2010 Active vibration control of piezoelectric intelligent structures, *Journal of Computers* **5** issue 3 401–09.

[4] Çimen T 2008 State-dependent Riccati equation (SDRE): A survey *Proceedings of the 17th World Congress of the IFAC, Seoul, Korea* 3761–75.

[5] Fenili A 2000 *Mathematical modeling and analysis of the ideal and nonideal behavior of slewing flexible structures* Ph.D Thesis, University of Campinas (UNICAMP), Faculty of Mechanical Engineering (Brazil) In portuguese.

[6] Abreu G L, Ribeiro J and Steffen Jr V 2003 Experiment on optimal vibration control of a flexible beam containing piezoelectric sensors and actuators *Shock and Vibration* 283–300.

[7] Abramowitz M and Stegun I (Eds.) 1972 *Handbook of mathematical functions with formulas, graphs, and mathematical tables* 9th printing (New York: Dover).

[8] LeVeque R J 2007 *Finite difference methods for ordinary and partial differential equations: steady-state and time-dependent problems* (Society for Industrial and Applied Mathematics) ISBN: 978-0-89871-629-0.

IOP Conf. Series: Journal of Physics: Conf. Series **1048** (2018) 012004 doi:10.1088/1742-6596/1048/1/012004

Mechanical response of a human femoral diaphyseal stabilized fracture using implant plate

I Herrera[1,4], B Rio[2,5], D Rodríguez[1,6] and R Lorente[3,7]

[1]Escuela de Ingenierías Industriales, Universidad de Extremadura, Avda. de Elvas s/n, 06006 Badajoz, Spain
[2]ETS de Ingenieros Industriales de Madrid. Departamento de Física Aplicada e Ingeniería de Materiales, Universidad Politécnica de Madrid, C/José Gutiérrez Abascal 2, 28006 Madrid, Spain
[3]SES Servicio Extremeño de Salud. Hospital Infanta Cristina. Avda. de Elvas s/n 06006 Badajoz Spain
[4]iherrera@unex.es, [5]bdelrio@etsii.upm.es, [6]daniel.tecno4@gmail.com, [7]rlorentem@hotmail.com

Abstract. The stabilization of a human femoral diaphyseal comminuted fracture using a fixation plate and screw system is analysed. A diaphyseal comminuted fracture represents a disruption in the medial femur with more than two bony fragments and bone loss. As a result of this, the osteosynthesis implant is subjected to a very high stress that could lead to its fracture, also the patient standing stability is affected due to higher slenderness of the lower limb. A procedure is proposed to simulate the surgery operation and the mechanical response of an in-vivo stabilized fracture. Stresses and displacements that occurred in the plate, fixation screws and bony fragments produced by the forces applied to the femoral head by the hip and to the diaphyseal cortex by tendons and muscles insertions during patient restoration are determined. Different fracture locations, cortical screws and fixation plates were analysed using the Finite Element Method which is configured to take into account heterogeneity and anisotropy of cortical and medullary bone. Recommendations about optimal implant design for the patient standing up stability and load bearing are concluded.

1. Introduction

The aim of this paper is to analyze different plates and screws combinations to stabilize a diaphyseal femur fracture, a 32-C3 diaphyseal femur fracture following the Müller AO Classification of Fractures—Long Bones [1], with a 30 mm gap between the upper and the lower part of the comminuted of this comminuted fracture, using the Finite Element Method. It was observed that the plate of several patients fractured unexpectedly several months after stabilization (See figures 1 and 2). This paper is intended to justify the phenomena and give criteria on patient stabilization procedures.

Three different fractured leg equivalent static load cases are of interest: $0.23{\cdot}BW$ vertical load exerted on the femur (BW is the body weight) for a patient using crutches carefully after operation (this is the maximum recommended, support on the fractured leg is prohibited), $0.75{\cdot}BW$ vertical load for a patient using crutches carelessly allowing partial standing on the fractured leg, and $1.643{\cdot}BW$ vertical load for the patient walking as normal (without crutches). This case is an extreme load case that should not happen in reality but it is of interest to estimate if the patient would suffer an injury or not.

Along with the forces due to the own weight of the patient on the femoral neck that is attached to the hip, forces due to the contraction of muscles attached to the femur should be considered. These muscles help the femur to support the weight of the patient and contract during walking, standing on one leg, climbing stairs and so on.

Unicortical and bicortical locking screws are used in different cases. The thread shape of the screws are not considered in this study. Several authors compare the screw thread effect on the simulation with that of considering the screw shaft and bone interaction as a bonded junction [2-4].

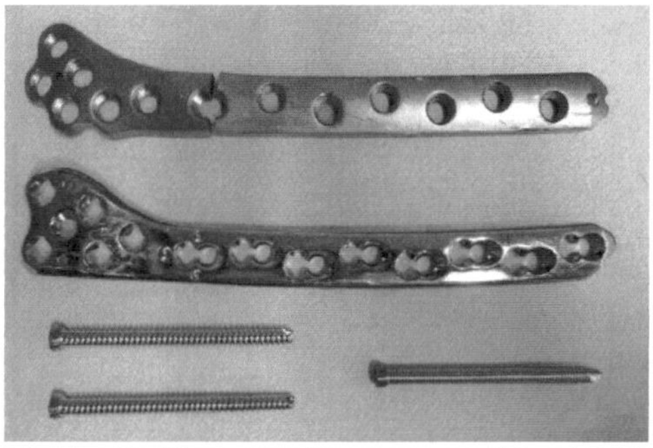

Figure 1. Plates and screws used to stabilize a 30mm long 32-C3 diaphyseal femur fracture. Upper (fractured) Lower (spare).

Figure 2. Fractured surface of the fractured plate shown in figure 1.

Normally, screws are not modelled as a helical 3D shape but as a 2D straight shape due to the complexity of the system and the high density mesh required to achieve realistic stresses on the plate and to get the highest stress on the neck of the screws zone. Bibliography about this point considers the high holding power of cortical screws (because of the very low thread pitch). Then the screw thread can be modelled as a smooth shaft and the screw-bone contact as a bonded junction. For the same reason the joint between the thread on the head of the screws and that on the hole of the plate can be modelled as a bonded junction too.

Implant plate fracture is initially motivated by fatigue during normal level walking of the patient. Shaat [5] has reported 316L stainless steel locking compression plate implants fatigue life based on femoral and tibial biomechanics during the gait.

During decades several authors have reported information about geometry and mechanical response of the human femur both by experiments and biomechanics modelling. Internal forces and moments on the medial femur during level walking activities were calculated mainly by equilibrium considerations and thigh isolation of the knee (Shaat[5], Taylor[6]) or thigh isolation of the hip (Duda[7], Bergmann [8]).

Brand [9] presented a mathematical model for predicting lower extremity muscle and joint forces based upon several cadavers. Duda [7] calculated internal forces and moments of the femur during gait taking into account all thigh muscles, body weight and contact forces. Taylor [6] measured *in-vivo* the forces in the distal femur and the knee during normal level walking activity using a distal femoral replacement. Tung-Wu[10] and Schneider[11] performed experiments on patients using a proximal femoral prostheses implanted after tumor resection and an intramedullary nail, respectively. In-vivo axial forces were calculated during level walking activity.

Seo[12] used the musculoskeletal simulation program Anybody Modeling System (Anybody Technology Inc. Denmark) and finite element analysis to calculate the stresses, strain and total deformation from a healthy male adult aged 29 (171cm height and 72kg weight) who walked normally.

IOP Conf. Series: Journal of Physics: Conf. Series **1048** (2018) 012004 doi:10.1088/1742-6596/1048/1/012004

2. The model

Let us first define the geometric model, materials properties and characteristics of the parts involved. The diaphysis (see figure 3) has an exterior layer of cortical type bone from 3 to 4 mm thick, whereas the bone interior is filled with trabecular or cancellous bone whose strength and holding power is about 10% that of the cortical bone [13] at the ends, and yellow bone marrow whose strength is null at the center. For simplicity, the 3D femur model includes only cortical bone material which occupies femur cortical bone space and occupied space by both yellow bone marrow and cancellous bone masses emptied. The cortical type screws used would stay in place due to the thread of the screw which is carving the cortical bone. Then the connection of the screws to the bone has been considered as a bonded junction.

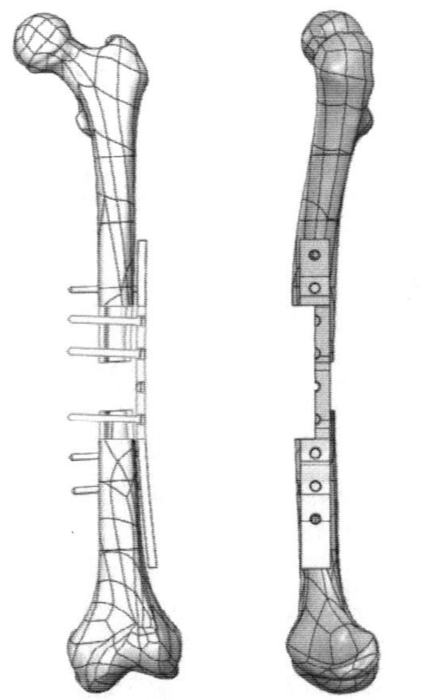

Figure 3. Sectional view of stabilized femur using fixation plate and bi-cortical screws.

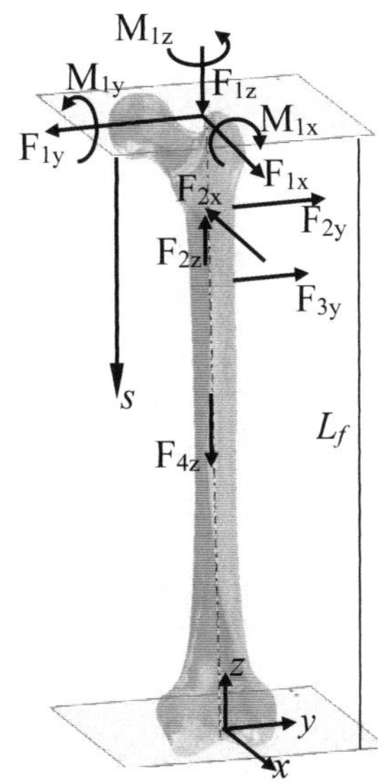

Figure 4. Schematic forces exerted on the left femur.

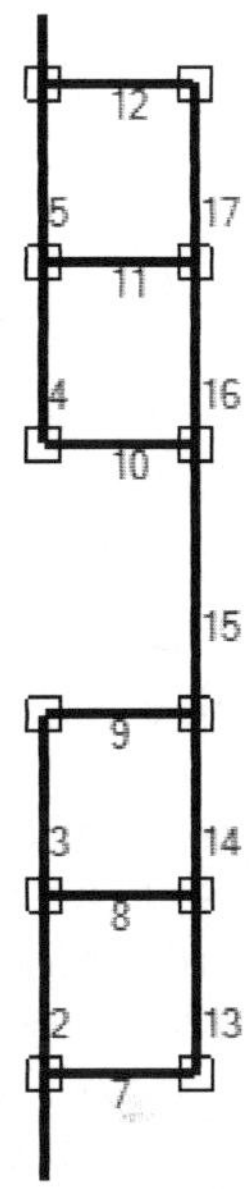

Figure 5. Truss bar model for stabilized femur using plate (13-17) and bi-cortical screws (7-12).

The femur shape was loaded from free internet library [14] and then scaled to get L_f*=404mm* length which corresponds to a *25-26%* of *1.63m,* the total height of a *58.5kg* weight patient. The femur geometry was modified to emptying the yellow bone marrow diaphyseal zone and complying with the above mentioned consideration about cancellous bone resistance (Figure 3). Diaphyseal ellipsoidal shape was minor axis *20.7mm* and major axis *24.6mm*. A procedure based in an artificial vision program is proposed to create the specific fractured femur geometry of a particular patient. A 3D finite element model can be built by scanning real human femur tomography images, subject the images to a number of filter processes and then apply a tool like "cover" of a drawing platform to build the 3D model.

The fixation plate studied (see figures 1 and 2) has threaded holes to lock the threaded screws head in, so it is a locking plate. During operation the traumatologist adapt the plate to the bone contour. Then, the plate is plastically deformed and bent to fit the approximate bone shape and then the gap between the bone and the plate is adjusted to a tolerance between *1mm* and *2mm* [2]. Similar process was followed to construct the 3D model from the bone and plate models (Figure 3). The deformed plate was constructed first using the Solidworks [15] sweep command to a directive curve which coincides with the nearest bone profile, separated *2mm* gap from the bone surface. The drill direction for every self-tapping screw is defined to be perpendicular to the directive curve. The cavity or similar command is applied to define a non-gap assembly between the plate and the screws and the screws and the bone.

The plate model is designed to simulate different fixation configurations by varying the plate thickness, the screws separation and type (compatible bi-cortical and uni-cortical fixation screws are feasible). The bicortical screws are *35mm* long and have a *4mm* shaft diameter. The head shape is threaded and shown in figure 1 which is similar to that shown at *ISO 5835:1991*.5835:1991. On the other hand, the uni-cortical screws will have the same shape than that of the bi-corticals but only *20mm* length. Two different plate widths are tested, *5.0mm* and *4.5mm*, both *200mm* length and *20mm* separation between threaded consecutive holes.

According to bibliography bone can be modelled as an elastic anisotropic material *120MPa* yield strength. In this study, bone will be considered orthotropic, *1850* kg/m^3 density and elastic mechanical properties [16] according to Table 1.

Table 1. Mechanical data for cortical type bone.

Young Modulus (MPa)	Shear Modulus (MPa)	Poisson coefficient
[1]E_1=16000	[2]G_{12}=3200	μ_{12}=0.30
E_2=6880	G_{23}=3600	μ_{23}=0.45
E_3=6300	G_{13}=3300	μ_{13}=0.30

[1]Subscript 1 refers to dyaphyseal femur longitudinal axis
[2]Subscript 2 refers to tangential direction and subscript 3 to radial direction.

The fractured fixation plate was characterized by micro-hardness measurements *310-330 HV0.2* in longitudinal section and *260-290HV0.2* in cross section, **SEM** (Scanning Electron Microscopy) and optic microscope test tube observation after Krolls reactive attack (see figure 6) showing light acridity compatible with cold stretching fabrication process and grain size *6* (*40-60*μm). The fixation plate was identified as commercially pure titanium *CP-Ti* in coarse grain condition which possesses an elastic limit $\sigma_e=155MPa$, endurance limit $\sigma_{f7}=120MPa$ $(10^7$ cycles), $\sigma_{f4}=280MPa$ $(1.5\cdot10^4$ cycles), tensile strength $\sigma_t=370MPa$ [17]. Typical elastic modulus and Poisson's ratio for *CP-Ti* are $E=105GPa$ and $\mu=0.3$, respectively.

The locking screws were treated as *Ti 6Al-4V* titanium alloy which possess an elastic modulus $E=104.8GPa$ and yield strength $\sigma_e=827MPa$

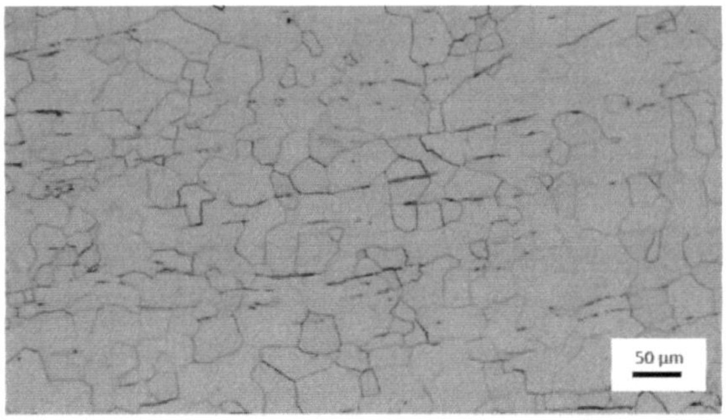

Figure 6. Optic micrograph for the fractured plate test tube after Krolls reactive attack showing light acridity and grain size *6*.

Internal moments and forces in the human femur during normal level walking has been estimated by different researchers (Duda[7], Taylor[6], Bergmann[8], Shaat[5]). The maximum healthful femur stress and displacements depend mainly on the vertical load exerted instantaneously on the thigh during the gait cycle and section location. A wide spread of data was found due to the extreme complexity of boundary conditions and high number of muscles needed on biomechanics models and scarce number of experiments which results are not directly extrapolable due to the inherent mechanical changes induced by the implants or prostheses in the thigh. This data is summarized in Table 2 below.

Table 2. Internal forces and moments exerted on femur. (*BW* is the body weight of the patient).

Simulation or Experiment configuration	[1] M_x [2] $s=0$ $[BWm]$	[1] M_x [2] $s=L_f$ $[BWm]$	[1] M_y [2] $s=0$ $[BWm]$	[1] M_y [2] $s=L_f$ $[BWm]$	[1] M_z [2] $s=0$ $[BWm]$	[1] M_z [2] $s=L_f$ $[BWm]$	[1] N [2] $s=0$ $[BW]$	[1] N [2] $s=L_f$ $[BW]$
Duda 10% gait cycle (Fig.4) All muscles	0.1300	0.2600	0.0011	0.2800	0.0055	0.0022	1.643	1.429
Duda 10% gait cycle (Fig.5) All muscles	0.1645	0.0188	0.0090	0.0380	0.1108	0.0386	1.955	1.610
Duda 20% gait cycle (Fig.5) All muscles	0.2250	0.0375	0.0190	0.0190	0.1060	0.0386	2.415	2.070
Duda 10% gait cycle (Fig.4) Hip muscles	0.1300	0.0011	0.0011	0.0022	0.0055	0.0055	1.643	1.643
Duda 10% gait cycle (Fig.4) Only glutei	0.1789	0.0800	0.0210	0.1050	0.0055	0.0055	1.643	1.250
Duda 10% gait cycle. Only Hip contacts	0.1789	0.2526	0.0210	0.2842	0.0400	0.0400	1.643	1.643
Bergmann (Hip implant Fig. 5 and 9)	0.0470	-	0.0080	-	0.0133	-	2.150	-
Taylor knee-femur implant at $s=0.5 \cdot L_f$	-	0.0718	-	0.0350	-	0.0115	-	1.860
Tung-Wu intramedullary nail at $s=0.38 \cdot L_f$							2.200	
Schneider intramedullary nail partial weighting 250N at $s \approx 0.5 \cdot L_f$	-	0.0720	-	0.0320	-	0.0880	-	1.320

[1] N is the compression force. M_x, M_y and M_z are the internal moments on the *x, y* and *z* axis respectively.

[2] s is the distance from the hip femur section to the specified section (See figure 4).

The forces exerted on the femur were calculated considering the internal forces and moments distribution due to body weight and contact forces at the hip and contact forces exerted on the diaphysis boundary by all thigh muscles at 10% gait cycle in normal level walking of the patient (Duda[7]). The gait cycle is between strike of the left leg and next heel strike of the same leg. The magnitude of forces modelled are summarized in table 3.

Table 3. Magnitude of forces exerted on the femur.

Force	Description	Value[1]
F_{1x}	Internal force at the femur head in *x* direction due to hip contact force and force exerted by gluteus maximus, gluteus medius, gluteus minimus and tensor fascia latae	$0.786 \cdot BW$
F_{1y}	Internal force at the femur head in *y* direction due to hip contact force and force exerted by gluteus maximus, gluteus medius, gluteus minimus and tensor fascia latae	$0.835 \cdot BW$
F_{1z}	Internal force at the femur head in *z* direction due to hip contact force and force exerted by gluteus maximus, gluteus medius, gluteus minimus and tensor fascia latae	$1.643 \cdot BW$
M_{1x}	Internal moment at the femur head in *x* direction due to hip contact force and force exerted by gluteus maximus, gluteus medius, gluteus minimus and tensor fascia latae	$0.1300 \cdot BWm$
M_{1y}	Internal moment at the femur head in *y* direction due to hip contact force and force exerted by gluteus maximus, gluteus medius, gluteus minimus and tensor fascia latae	$0.0011 \cdot BWm$
M_{1z}	Internal moment at the femur head in *z* direction due to hip contact force and force exerted by gluteus maximus, gluteus medius, gluteus minimus and tensor fascia latae	$0.0055 \cdot BWm$
F_{2x}	Force exerted by psoas major, iliacus, piriformis, obturatori, gemelli, popliteus, quadratus, pectineus and vastus lateralis at $s=0.15 \cdot L_f$ in *x* direction	$0.926 \cdot BW$
F_{2y}	Force exerted by psoas major, iliacus, piriformis, obturatori, gemelli, popliteus, quadratus, pectineus and vastus lateralis at $s=0.15 \cdot L_f$ in *y* direction	$0.497 \cdot BW$
F_{2z}	Force exerted by psoas major, iliacus, piriformis, obturatori, gemelli, popliteus, quadratus and pectineus and vastus lateralis at $s=0.15 \cdot L_f$ in *z* direction	$0.572 \cdot BW$
F_{3y}	Force exerted by gluteus maximus, adductor brevis and biceps femoris breve at $s=0.35 \cdot L_f$ from femur head in *y* direction	$0.1427 \cdot BW$
F_{4z}	Force exerted by adductor longus, biceps femoris breve, vastus medialis and vastus intermedius at $s=0.55 \cdot L_f$ from femur head in *z* direction	$0.3575 \cdot BW$

[1]Direction of forces and moments are shown in figure 4.

IOP Conf. Series: Journal of Physics: Conf. Series **1048** (2018) 012004 doi:10.1088/1742-6596/1048/1/012004

F_{1y}, F_{2x}, F_{2y}, F_{2z}, F_{3y} and F_{4z} forces (figure 4) has been determined by equilibrium considerations from the *x*, *y*, *z* internal forces and *x*, *y* internal moments diagrams of the femur shown by Duda [7].

Following previous screw-plate and screw-bone connections hypothesis, an equivalent 2D truss bar structure of the sane femur and stabilized femur (averaged *25mm* diameter *4mm* thickness and isotropic material with elastic modulus $E=16GPa$ cortical bone) can be derived (see figure 5).

In addition, FEA simulations were performed in order to get estimates of the stress distribution on femur, plate and screws. FEA modelling was implemented in the SolidWorks platform using small displacement and small strain solver [15]. See figure 3. *76998* tetraedric elements compose a mesh with a minimum size of 2.*8mm* and maximum of *13.9mm* on non-critical regions like those of the lower end of the femur. Various mesh densities and mesh control were simulated to confirm that the number of elements were enough so the displacement results will not vary significantly due to meshing.

Compatible stress calculations of the structural system shown in figure 4 demands that the tibio-femoral joint were considered a fixed joint and that the hip joint of the femoral head to the acetabulum of the hip were considered to move free. Femur displacements are measured with respect to the straight line connecting both deformed ends. These displacements can be estimated from the quarter of the fixed-free ends displacements considering similarity with the pinned ends bar deflections.

3. Results and discussion

Several stabilization configurations were performed varying the number and type of stabilization screws and fracture location using both 2D truss bar model and FEA model. The denomination of a particular configuration consists in six figures: first three figures for the type of screw used to thread in every hole of the three possible in the upper femur followed by another three figures for those of the lower femur holes. Any figure can be: ***B*** for a bicortical screw inserted in the hole, ***U*** for unicortical screw or "-" for free hole.

Table 4 below shows the femur, screws and plate maximum sectional normal stresses in the direction coincident with the longitudinal axis of the femur (this is a good estimate of Von Mises equivalent stress) and system maximum displacement (expressed as displacement resultant for the 2D truss bar).

Table 4. Data retrieved from 2D truss structure analysis (*BW=650N*)

Configuration	Width Plate (mm)	0.230·BW load Plate/Screw/Bone Stress [MPa]	0.230·BW load Displ. [mm]	0.750·BW load Plate/Screw/Bone Stress [MPa]	0.750·BW load Displ. [mm]	1.643·BW load Plate/Screw/Bone Stress [MPa]	1.643·BW load Displ. [mm]
BBBBBB s=0.39·L$_f$	5.0	53/189/8	0.47	174/616/27	1.54	380/1350/60	[1]3.38
	4.5	64/186/8	0.92	210/607/27	1.79	459/1330/60	3.93
BBBBBB s=0.49·L$_f$	5.0	41/176/7	0.45	133/575/23	1.47	292/1260/51	3.23
	4.5	49/174/7	0.86	159/566/23	1.68	348/1240/51	3.68
BBBBBB s=0.59·L$_f$	5.0	25/168/6	0.36	82/548/18	1.16	179/1200/40	2.55
	4.5	29/164/6	0.39	96/534/18	1.26	210/1170/40	2.75
B-BB-B s=0.59·L$_f$	5.0	27/217/6	0.38	89/707/20	1.22	195/1550/43	2.68
	4.5	32/216/6	0.41	105/703/20	1.32	230/1540/43	2.90
UBBBBU s=0.59·L$_f$	5.0	25/190/6	0.47	82/621/20	1.53	179/1360/43	3.35
	4.5	32/186/6	0.49	103/607/20	1.60	226/1330/43	3.50
-BBBB- s=0.59·L$_f$	5.0	28/284/6	0.61	91/927/20	1.98	200/2030/43	4.33
	4.5	33/283/6	0.62	109/922/20	2.03	238/2020/43	4.45
BBBBBB s=0.69·L$_f$	5.0	15/167/5	0.30	49/543/16	0.97	105/1190/36	2.13
	4.5	17/153/5	0.30	56/498/16	0.98	123/1090/36	2.15
No fractured femur		7		22		48	1.83

[1]Figures in red means exceeded value with respect to that of healthful femur performance or component stress that exceeds its elastic limit.

If we focus on the plate stresses results, it must be observed that 2D structure analysis uses Strength of Materials formulas for bars to get the maximum sectional normal stress. In critical regions, the maximum stress could be, at least, *2.5* times that of the maximum normal stress shown in Table 4 due to stress concentration phenomena which could lead to very early plate fracture depending on the configuration. This would mean that only stresses below *110MPa* would be safe. This conducts to prohibit support on the damaged leg and only partial standing if the fracture location is farther than $0.49 \cdot L_f$ from hip femur head.

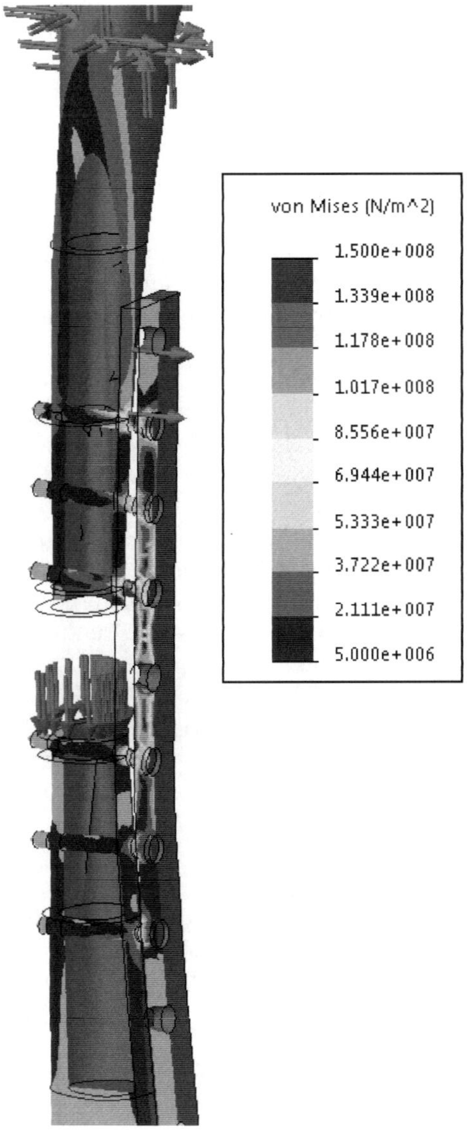

Figure 7. FEA results. ***BBBBBB*** stabilized femur fractured at $s=0.49 \cdot L_f$ subjected to $1.643 \cdot BW$ Vertical load ($BW=650N$).

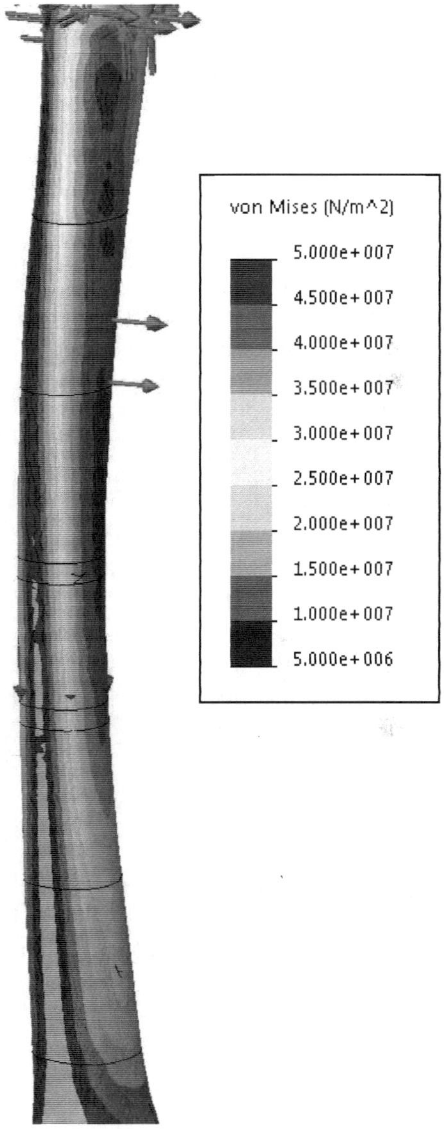

Figure 8. FEA results. Sane femur subjected to $1.643 \cdot BW$ Vertical load ($BW=650N$).

The difference between *5mm* or *4.5mm* plate thickness results are more pronounced on the displacements results that on the stresses. Though *5mm* plate thickness is more up to reduce bending stresses the centroid of its sections has to be separated an extra *0.25mm* from the femur increasing bending moments.

With respect to the current debate between making a fixation more or less rigid, several researchers conclude that a thinner plate make the bone support more stressed, the bone cures faster and prevents it for becoming atrophied due to Wolff's law [13]. Also, higher stresses near the screw holes of the bone can produce hypertrophy on those zones.

For some configurations, a small displacement small strain static simulation were performed according to the FEA model presented in previous paragraph. Solidworks platform [15] were used. The equivalent Von Mises stresses and magnitude of the maximum displacement were calculated. FEA results are shown in figures 7 and 8 for comparison of the sane femur stresses with stabilized femur stresses. Good agreement was found with 2D Structural Analysis displacements but considerably higher stresses due to stress concentration in the critical regions. FEA analysis reveals that maximum overall stress appears in the nearest holes of the plate to its center (even free hole), in the gap fragment of the screws due to tangent force and in the holes practiced by the traumotologist on the femur fragments.

Though FEA analysis is a good tool for stress concentration problems it was found that it did not provide a good distribution of results if the mesh was increased in the critical regions, some convergence fault inherent to the method appears. Then, no quantitative information on stresses could be discussed and future research should be done on this point.

If we get back to table 4 and compare stresses and displacements for the three load cases, it is clear that the recommended use of crutches during patient restoration, even allowing partial standing, is safe for all the screws configurations except for ***–BBBB-*** configuration which would be the case of weak fixation of the screw on the plate or reduced femur fragment space. If we take into account repetitive load and unload which is the normal walking case, additional fatigue analysis will be essential to allow partial standing of the patient, the application of other configurations different to the safer ***BBBBBB*** configuration or the change for a best performance material or component dimensions.

With respect to fracture location proximity to femur head is more critical. With respect to the number of screws for fixing one femur fragment comparing ***BBBBBB*** with ***B-BB-B*** and ***–BBBB-*** reveals that three screws reduce the screw stresses considerably. If only two screws are possible it is preferred to locate them as far as possible (***B-BB-B*** is preferred to ***–BBBB-***, for example).

An alternative for a patient could walk without crutches, stresses results reveals that it is the screws the weak point of the system. This could be corrected increasing the screw diameter from *4mm* to *4.5mm* which is feasible but this need detailed analysis because this change affects the plate and bony fragments performance too.

4. Conclusions and future work

The stabilization of a human femoral diaphyseal comminuted fracture using a fixation plate and screw system has been modelled and simulated. A number of typical patient habits during restoration like walking as normal or the carelessly use of crutches are considered. Different stabilization configurations were performed varying the number and type of stabilization screws and using FEA modelling implemented in the SolidWorks platform.

The deformation process similar to that of the traumatologist when adapting the plate to the bone contour during operation was simulated to get realistic 3D model of a stabilized femur.

It was found that the mechanical characteristics and stress concentration factors in the critical regions of the fixation plate compromises the restoration period without using crutches and even partial standing on the fractured leg of the patient if the fracture is located close to the femur hip head.

Unexpected fractures of an in-vivo implant plate were analyzed. The model and simulation performed is simple and accurate enough to justify why the plate of a patient could fracture unexpectedly after stabilization if good estimates of the stress concentration factors in the critical regions of the plate are available. Mechanical fatigue of a Titanium alloy fixation plate is compatible with fracture after four patient restoration months and clinical data.

The model and procedure presented in this investigation when applied to a sample of similar patients' fractures could be useful to generalize conclusions. This could reduce the number of surgery operations due to the fracture of the fixation plate. Furthermore, the methodology proposed in the paper can be considered as a useful tool for improving patient stabilization procedures.

More precise estimates of the fatigue life of fixation plates requires a more precise determination of the stress concentration factor in the critical regions observed in this paper.

References

[1] Müller M, Nazarian S and Koch P 1987 *Classification AO des Fractures* (New York: Springer-Verlag) Long Bones

[2] Ahmad M, Nanda R, Bajwa A, Candal-Couto J, Green S and Hui A 2007 Biomechanical testing of the locking compression plate: when does the distance between bone and implant significantly reduce construct stability *Injury, Int. J. Care Injured* **38** 358—364.

[3] Chatzistergos P, Magnissalis E and Kourkoulis S 2010 A parametric study of cylindrical pedicle screw design implications on the pullout performance using an experimentally validated finite-element model *Medical Engineering & Physics* **32** 145–154.

[4] Feerick E and McGarry P 2012 Cortical bone failure mechanisms during screw pullout. *Journal of Biomechanics* **45** 1666–1672.

[5] M Shaat Reporting the Fatigue Life of Life of 316L Stainless Steel Locking Compression Plate Implants: The Role of the Femoral and Tibial Biomechanics During the Gait 2017 *Journal of Biomechanical Engineering* **139** 104502-1-5

[6] Taylor S, Walker P, Perry J, Cannon S, FRCS and Woledge R 1998 The forces in the distal femur and the knee during walking and other activities measured by telemetry *The Journal of Arthroplasty* **13** 4 428-436

[7] Duda G, Schneider E and Chao E 1997 Internal Forces and Moments in the Femur during walking *J. Biomechanics* **30** 9 933-941

[8] Bergmann G, Deuretzbacher G, Heller M, Graichen F, Rohlmann A, Strauss J and Duda G 2001 *Journal of Biomechanics* **34** 859-871

[9] Brand R, Crowninshield R, Wittstock C, Pedersen D, Clark C and Krieken F 1982 A model of lower extremity muscular anatomy *Transactions of the ASME* **104** 304-310

[10] Tung-Wu L, Taylor S, O'Connor J and Walker P 1997 Influence of muscle activity on the forces in the femur: an in vivo study *J. Biomechanics* **30** 11,12 1101-1106

[11] Schneider E, Michel M, Genge M, Zuber K, Ganz R and Perren S 2001 Loads acting in an intramedullary nail during fracture healing in the human femur *Journal of Biomechanics* **34** 849-857

[12] Seo J, Kanga D, Kim J, Yang S, Kim D, Choia J and Tacka G 2014 Finite element analysis of the femur during stance phase of gait based on musculoskeletal model simulation *Bio-Medical Materials and Engineering* **24** 2485–2493 DOI 10.3233/BME-141062 IOS Press

[13] Nordin M and Frankel V M *Basic Biomechanics of the Musculoskeletal System.* 3rd edition. Lippincoit Williams & Wilkins.

[14] Redding Engineering LLC 2008 User library-large human femur made from CT image scan and surfaced in Solidworks. www.3dcontentcentral.com

[15] SolidworksTM 2016 Dassault Systémes.

[16] Guerrero V Luis A, Ramírez D Edgar I, Ruiz C, Osvaldo, Ortiz P Armando 2015 Metodología del modelado mediante FEM de un fémur humano y su análisis bajo distintas condiciones in vivo. *Memorias del XXI Congreso Internacional anual de la SOMIM, Coatzacoalcos, Veracruz, México* 0495-0503.

[17] Alkhazraji H, El-Danaf E, Wollmann M and Wagner L 2015 Enhanced fatigue strength of commercially pure Ti processed by rotary swaging *Advances in Materials Science and Engineering* **2015** ID 301837, 12 pages http://dx.doi.org/10.1155/2015/301837

The modelling and prediction of dynamic responses of slender continua deployed in tall structures under long-period seismic excitations

S Kaczmarczyk

Faculty of Arts, Science and Technology, University of Northampton, St. George's Avenue, Northampton NN2 6JD, UK

stefan.kaczmarczyk@northampton.ac.uk

Abstract. Tall building structures are susceptible to large sway motions when subjected to earthquake excitations. They are particularly affected by long period earthquake ground motions. These low frequency seismic waves resonate with the fundamental mode of the building structure which in turn causes resonance interactions with long slender continua deployed in modular non-structural installations, such as lifts. Damage due to large resonance motions of suspension/ compensating ropes and cables during earthquake are one of the most common modes of failure in high-rise lift installations. In this paper an analytical model to predict the dynamic responses of a suspension rope system installed in a tall host structure under seismic conditions is presented. The model is then used to predict the dynamic performance of the system under earthquake excitations. The predictions can then be used to develop suitable mitigating strategies and protective measures to minimize the earthquake damage.

1. Introduction

Long-period ground motion excitations induced by distant earthquakes can result in resonance and large seismic responses of high-rise buildings in the modern mega cities located in the intra-plate regions such as Hong Kong, Shanghai, New York, Singapore and Dubai [1]. In those conditions severe damage is inflicted to the buildings' structure and/or to their non-structural components such as lifts and escalators [2,3]. In this paper an analytical model to predict the dynamic responses of a cable - mass system which represents a lift suspension rope – car/ counterweight system under seismic conditions is presented. The model is then used to predict the dynamic performance of the system under long-period earthquake excitations.

2. Mathematical model

Figure 1 shows a cable - mass system mounted within a vertical cantilever host structure subject to ground motion $s_0(t)$. The mass M is suspended on the cable of length L and is constrained horizontally within the host structure by a spring – viscous damping element of effective coefficient of stiffness k the coefficient of damping c. The upper end of the cable is passing through O and the height of the structure is $AB = Z_0$. The structure undergoes bending elastic deformations $\overline{w}(z,t)$ where $0 \le z \le Z_0$, with the displacements at the top end defined as $\overline{w}(Z_0,t) = \overline{w}_0(t)$. The cable - mass - spring system moves vertically within the host structure at transport speed *V* and acceleration *a*. The mean quasi-static tension, mass per unit length, modulus of elasticity and cross-sectional area of the cable are denoted as $T^i = [M + m(L-x)](g-a)$, *m, E* and *A*, respectively. The spatial coordinate *x* is measured from the upper end of the cable downwards as shown. The lateral dynamic displacements of the cable, coupled with the longitudinal displacements *u(x,t),* are denoted as *v(x,t).* The lateral and longitudinal motions of mass *M* are shown as $v_M(t)$ and $u_M(t)$, respectively.

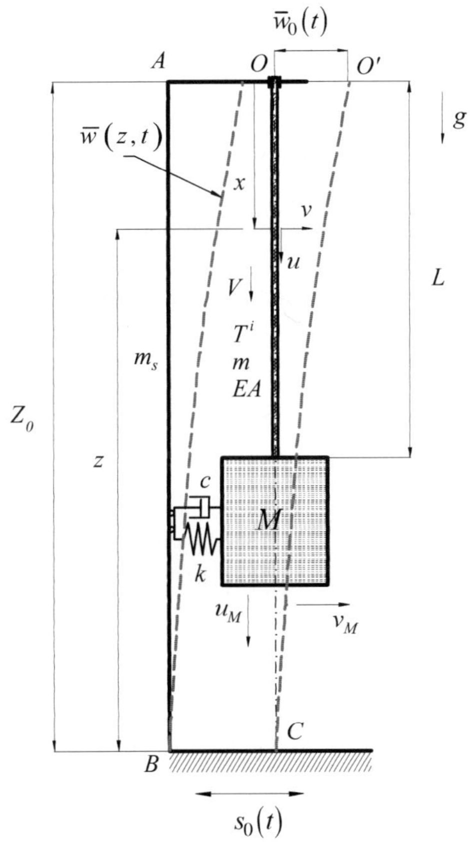

Figure 1. Cable – mass system.

2.1. *Model of the structure*

The dynamic response of the structure is introduced as the base motion excitation acting upon the cable - mass - spring system. The overall displacements $w(z,t)$ of the structure are described by the following equation

$$m_s(z)w_{tt} + \mathcal{C}w_t + \mathcal{L}w = 0 \tag{1}$$

where $m_s(z)$ the linear mass density of the structure, $\mathcal{L}$ is the spatial operator related to the elastic potential energy of the structure, C is the damping operator, and $(\ )_t$ denotes the partial derivative with respect to time *t*. The overall displacements of the structure are expressed as

$$w(z,t)=\overline{w}(z,t)+s_0(t) \tag{2}$$

Equation (1) can then be expressed as

$$m_s(z)\overline{w}_{tt}+C\overline{w}_t+\mathcal{L}\overline{w}=-m_s(z)\ddot{s}_0(t) \tag{3}$$

The solution of (3) can then be assumed in the form of modal expansion

$$\overline{w}(z,t)=\sum_{n=1}^{\infty}W_n(z)p_n(t) \tag{4}$$

where $W_n(z)$ are the eigenfunctions of the structure and $p_n(t)$ are the natural (modal) coordinates. Using equation (4) in equation (3), multiplying the result by W_r integrating over the domain $0<z<Z_0$ and considering the eigenfunction orthogonality conditions the following modal equations are obtained

$$\ddot{p}_r(t)+2\zeta_r\omega_r\dot{p}_r(t)+\omega_r^2p_r(t)=P_r(t) \tag{5}$$

where ω_r are the natural frequencies, ζ_r are the modal damping ratios and $P_r(t)$ are the modal excitation terms defined as

$$P_r(t)=-\frac{\ddot{s}_0(t)}{m_r}\int_0^{Z_0}m_s(z)W_r(z)dz \tag{6}$$

where $m_r=\int_0^{Z_0}m_s(z)W_r^2(z)dz$ represents the modal mass. The steady-state response of the structure can then be determined from equation (5) as

$$p_r(t)=\frac{1}{\omega_{dr}}\int_0^t P_r(t-\tau)e^{-\zeta_r\omega_r\tau}\sin\omega_{dr}\tau d\tau \tag{7}$$

where $\omega_{dr}=\omega_r\sqrt{1-\zeta^2}$. If the eigenvalue problem associated with equation (3) does not admit a closed-form solution a set of comparison functions can be used to seek an approximate solution of equation (3) in the following form [4]

$$\overline{w}(z,t)=\sum_{n=1}^{N}\Psi_n(z)q_n(t) \tag{8}$$

where $q_n(t)$ are generalised coordinates. Using equation (8) in equation (3) leads to the following matrix equation of motion

$$\boldsymbol{M}\ddot{\boldsymbol{q}}+\boldsymbol{C}\dot{\boldsymbol{q}}+\boldsymbol{K}\boldsymbol{q}=\boldsymbol{F}(t) \tag{9}$$

where $\boldsymbol{M}=[m_{rn}],\ m_{rn}=\int_0^{Z_0} m_s\Psi_r\Psi_n dz,\ \ \boldsymbol{C}=[c_{rn}],\ c_{rn}=\int_0^{Z_0}\Psi_r C\Psi_n dz\ \ \boldsymbol{K}=[k_{rn}],\ k_{rn}=\int_0^{Z_0}\Psi_r\mathcal{L}\Psi_n dz$,

$r,n=1,\ldots N$, and $\boldsymbol{F}(t)=[F_r(t)],\ F_r(t)=-\ddot{s}_0(t)\int_0^{Z_0} m_s\Psi_r dz$ is a vector of generalized excitation forces.

2.2. Governing equations of the cable - mass system

Consider the motions of the cable - mass system. The equations governing the dynamic response of the system are developed by the application of the Hamilton's principle which requires that

$$\int_{t_1}^{t_2}(\delta E-\delta\Pi-\delta W_{nc})dt=0,\ \ \delta w(x,t)=0 \text{ at } t=t_1,t_2 \tag{10}$$

where E, Π and W_{nc} denote the kinetic energy, the potential energy and the work due to non-conservative forces acting upon the system, respectively. The kinetic energy is given as

$$E=\int_0^L \hat{E}_m(\dot{u},\dot{v})dx+E_M(\dot{u}_M,\dot{v}_M) \tag{11}$$

where

$$\hat{E}_m=\frac{1}{2}m\left[(\dot{u}+V)^2+\dot{v}_M\right] \tag{12}$$

is the kinetic energy density of the cable, and

$$E_M=\frac{1}{2}M\left[(\dot{u}_M+V)^2+\dot{v}_M^2\right] \tag{13}$$

is the kinetic energy of the end mass. Here $u_M=u(L,t)$, $v_M=v(L,t)$, overdots denote the total derivatives with respect to time expressed as $\dot{u}\equiv\frac{Du}{Dt}=u_t+Vu_x$ and $\dot{v}\equiv\frac{Dv}{Dt}=v_t+Vv_x$, with $(\ )_x$ denoting partial derivatives with respect to x, respectively. The potential energy is given as

$$\Pi=\int_0^L\left[\hat{\Pi}_e(u_x,v_x)-mgu\right]dx-Mgu_M+\frac{1}{2}k\Delta^2 \tag{14}$$

where it is k is assumed to be constant and represents the flexibility of the mass guiding system (in the context of a lift system this would be the flexibility of the car roller guide system, the guide rail – guide rail/ bracket system can be considered as rigid). Furthermore the elastic potential energy density of the cable $\hat{\Pi}_e$ is expressed in terms of the Green's strain measure ε as

$$\hat{\Pi}_e=\left(T^i+\frac{1}{2}AE\varepsilon\right)\varepsilon \tag{15}$$

where the Green's strain measure is defined as [5]

$$\varepsilon=u_x+\frac{1}{2}v_x^2 \tag{16}$$

The last term in equation (14) represents the potential energy of the spring deformed by $\Delta=v_M-w_M$ where $w_M=w(Z_0-L,t)$. The work done by the non-conservative damping force is expressed as

$\delta W_{nc} = -c\dot{\Delta}\delta v_M$ so that Hamilton's principle yields the following partial differential equations of motion

$$
\begin{aligned}
&m\frac{D^2u}{Dt^2} - EA\varepsilon_x = 0, \\
&m\frac{D^2v}{Dt^2} - Tv_{xx} + m(g-a)(xv_{xx}+v_x) - EA(\varepsilon v_x)_x = 0; \\
&M\ddot{v}_M + T^i(L)v_x\big|_{x=L} + c\dot{\Delta} + k\Delta + EA\varepsilon\big|_{x=L} v_x\big|_{x=L} = 0; \\
&M\ddot{u}_M + EA\varepsilon\big|_{x=L} = 0
\end{aligned} \tag{17}
$$

where $T = (M + mL)(g - a)$. For tensioned members such as metallic cables the lateral frequencies are much lower than the longitudinal frequencies. Thus, considering that in the case of long-period ground motions the excitations frequencies are much lower than the fundamental longitudinal frequencies the longitudinal inertia of the cable can be neglected in the first equation in (17) [5]. After integrating this equation and using the boundary conditions $u(0,t) = 0$, $u(L,t) = u_M$, the following expression for the quasi-static axial strain in the cable results

$$\varepsilon = u_x + \frac{1}{2}v_x^2 = \frac{u_M}{L} + \frac{1}{2L}\int_0^L v_x^2 dx \equiv e(t) \tag{18}$$

In that way the dynamic model (17) of the system is reduced to three equations of motion.

2.3. Base excitation

In order to develop the model further the assumption is made that the influence of the cable system dynamics on the structural response can be neglected. By considering the scenario in which the structure is subject to fundamental resonance an approximate solution of equation (3) can be sought by using a single-term expansion with the polynomial shape function $\Psi(\eta) = 3\eta^2 - 2\eta^3$ where $\eta = z/Z_0$, applied in equation (8). The base motion excitation is then introduced by expressing the overall lateral displacements of the cable – mass as

$$v(x,t) = \bar{v}(x,t) + s_0(t) + \left(1 + \frac{\Psi_L - 1}{L(\tau)}x\right)\bar{w}_0(t) \tag{19}$$

where $\Psi_L = \Psi\left(\frac{Z_0 - L(\tau)}{Z_0}\right)$ and $\bar{w}_0(t) = \bar{w}(Z_0,t)$. In this formultion L is considered a slowly varying function in time meaning that the change of lenght L over a period corresponding to the fundamental frequency of the system is small compared to the instantaneous value of L [6]. In order to represent this fact a slow time scale defined as $\tau = \epsilon t$, where $\epsilon \ll 1$ is a small parameter, has been introduced. The first term in equation (19) is then expressed

$$\bar{v}(x,t) = \sum_{n=1}^{N} \Phi_n\left[x; L(\tau)\right] q_n(t) \tag{20}$$

where Φ_n are orthogonal trial functions given as

$$\Phi_n\left[x; L(\tau)\right] = \sin\left[\sigma_n\left(L(\tau)\right)x\right], \quad n = 1,2,\ldots,N \tag{21}$$

The trial functions (21) are defined in terms of the slowly varying eigenvalues σ_n determined by the frequency equation

$$\left(k-\frac{M}{m}T_M\sigma_n^2\right)\sin(\sigma_n L)+T_M\sigma_n\cos(\sigma_n L)=0,\ T_M=T^{i}(L) \tag{22}$$

By using equation (19) in the reduced system of three equations obtained from (17), and applying the expansion (20) the following set of ordinary differential equations results

$$\begin{gathered}\ddot{q}_r+2\tilde{\zeta}_r\tilde{\omega}_r\dot{q}_r+\sum_{n=1}^{N}C_{rn}q_n+\tilde{\omega}_r^2q_r+\sum_{n=1}^{N}\left[K_{rn}+\frac{EA}{\tilde{m}_r}\left(\frac{\Psi_L-1}{L}\overline{w}_0\right)^2\left(\frac{1}{L}A_{rn}-\frac{1}{2}\Gamma_{rn}\right)\right]q_n+\frac{EA}{\tilde{m}_rL}\frac{\Psi_L-1}{L}\alpha_r\overline{w}_0u_M\\ =\frac{EA}{\tilde{m}_rL}\left[\sum_{n=1}^{N}\Gamma_{rn}q_nu_M+\frac{1}{2}\frac{\Psi_L-1}{L}\overline{w}_0\sum_{i=1}^{N}\sum_{j=1}^{N}\left(2\alpha_i\Gamma_{rj}-\alpha_r\kappa_{ij}\right)q_iq_j+\frac{1}{2}\sum_{i=1}^{N}\sum_{j=1}^{N}\sum_{k=1}^{N}\Gamma_{rk}q_iq_jq_k\right]+\tilde{Q}_r(t;\tau)\\ \ddot{u}_M+2\zeta_M\omega_M\dot{u}_M+\omega_M^2u_M+\frac{EA}{ML}\frac{\Psi_L-1}{L}\overline{w}_0\sum_{n=1}^{N}\alpha_nq_n=-\frac{1}{2}\frac{EA}{ML}\sum_{i=1}^{N}\sum_{j=1}^{N}\kappa_{ij}q_iq_j+\tilde{Q}_u(t;\tau)\end{gathered} \tag{23}$$

where $\tilde{\zeta}_r$ and ζ_M reperesent the damping ratios and

$$\begin{gathered}\tilde{\omega}_r^2=\sigma_r^2\frac{T_M}{m};\ \alpha_r=\Phi_r(L),\ \tilde{m}_r=\int_0^L\Phi_r^2dx+M\alpha_r^2,\ A_{rn}=\alpha_r\alpha_n\\ K_{rn}=\frac{m}{\tilde{m}_r}\left[g\Psi_{rn}+\left[V^2-L(g-a)\right]\Upsilon_{rn}+(g-a)\Theta_{rn}\right],\ C_{rn}=2\frac{m}{\tilde{m}_r}V\Psi_{rn}+\frac{1}{\tilde{m}_r}cA_{rn}\\ \Theta_{rn}=\int_0^L x\Phi_n''\Phi_rdx,\ \Psi_{rn}=\int_0^L\Phi_n'\Phi_rdx,\ \Upsilon_{rn}=\int_0^L\Phi_n''\Phi_rdx,\ \kappa_{ij}=\int_0^L\Phi_i'\Phi_j'dx,\ \Gamma_{rn}=\Upsilon_{rn}-\alpha_r\Phi_n'(L)\\ \tilde{Q}_r(t;\tau)=\frac{1}{\tilde{m}_r}\left[Q_r(t;\tau)-\frac{EA}{2}\alpha_r\left(\frac{\Psi_L-1}{L}\overline{w}_0(t)\right)^3\right];\\ Q_r(t;\tau)=-(mX_r+M\alpha_r)\ddot{s}_0-\left[m\left(X_r+\frac{\Psi_L-1}{L}\Pi_r\right)+M\Psi_L\alpha_r\right]\ddot{\overline{w}}_0-2mV\frac{\Psi_L-1}{L}X_r\dot{\overline{w}}_0+\\ -\frac{\Psi_L-1}{L}(mgX_r+T_M\alpha_r)\overline{w}_0;\ X_r=\int_0^L\Phi_rdx,\ \Pi_r=\int_0^Lx\Phi_rdx\\ \tilde{Q}_u(t;\tau)=-\frac{EA}{2M}\left(\frac{\Psi_L-1}{L}\overline{w}_0\right)^2\end{gathered} \tag{24}$$

3. Numerical example and results

A parametric study has been conducted for a model of the cable - mass system comprising the mass M = 3600 kg suspended on 6 steel wire ropes. The ropes have mass per unit length 0.872 kg/m and longitudinal stiffness EA = 22.889 MN, each. Figure 2 shows the variation of the first four lateral natural frequencies of the system when the length of the ropes changes from L_{min} = 58.66 m to L_{max} = 258.66 m. The red solid lines represent the frequencies when the spring constant is k = 66.698 kN/m. The black dashed lines represent the natural frequencies corresponding to the spring of coefficient of elasticity of one order of magnitude higher (k = 666.98 kN/m). In the system with the softer spring the curve veering phenomena can be observed when two eigenvalues approach each other closely and suddenly veer away [7]. For example, the veering regions of the 2nd and the 3rd natural frequency loci and the 3rd and the 4th natural frequency loci correspond to the lengths L of about 125 m and about 193 m, respectively. It is also interesting to observe that the flat sections of the 2nd, 3rd and 4th natural frequency curves correspond to the modes of the mass M motion. The frequency value can be estimated as $\overline{\omega}_M=\sqrt{k/M}=4.3043$ rad/s ($\overline{f}_M=\overline{\omega}_M/2\pi=0.6851$ Hz, shown as the green horizontal dashed line in figure 2). The first/ fundamental longitudinal natural frequency, corresponding to the frequency $\omega_M = EA/ML$ is also shown in figure 1 (solid blue line).

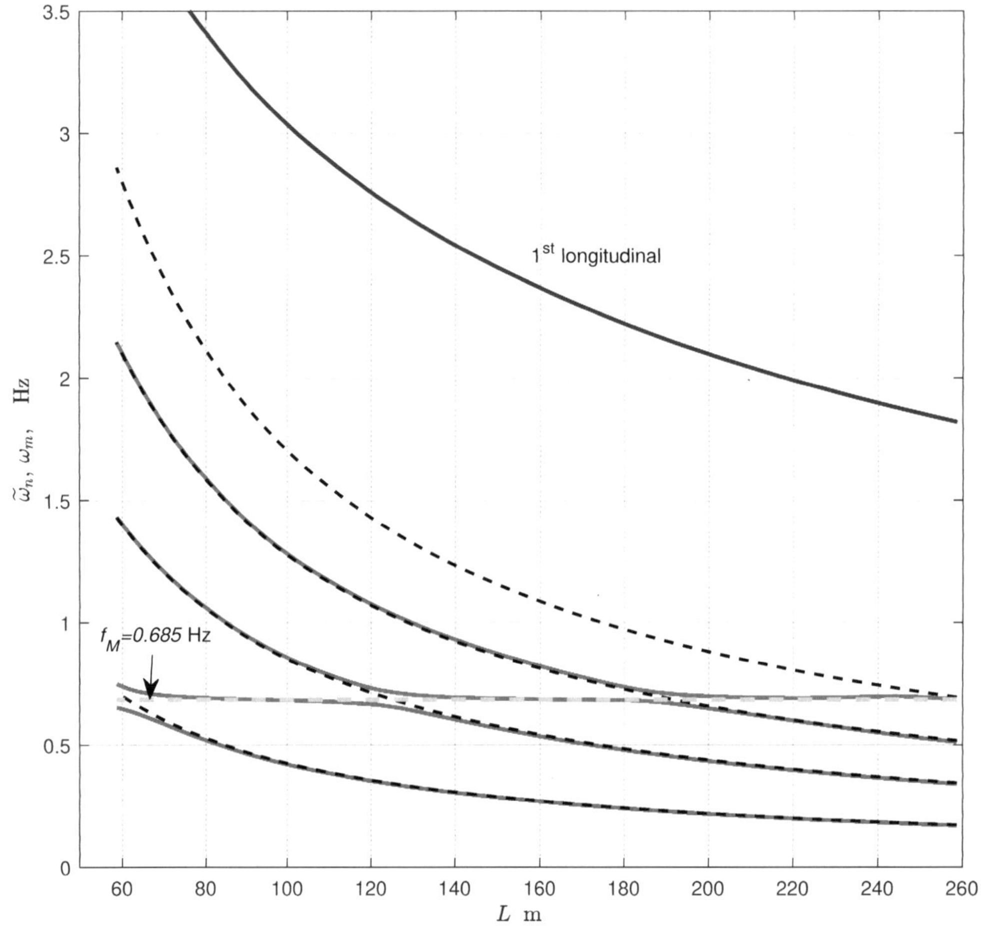

Figure 2. The natural lateral frequencies of the cable - mass system (red solid for k = 66.698 kN/m; black dashed lines for k = 666.98 kN/m), the frequency of the mass M ($\bar{\omega}_M = \sqrt{k/M}$ green solid line) and the 1st longitudinal frequency (blue solid line).

Figure 3 shows the lateral displacements $\bar{v}(x,t)$ of the cable – mass system when the mass is ascending slowly from the lower level upwards at the reduced speed of 0.75 m/s and the length of the ropes changes from about 140 m to 80 m, respectively. The dominant frequency of the ground motion is assumed to be 0.68 Hz (and near $f_M = 0.6851$ Hz), $s_0(t)$ is harmonic with the acceleration magnitude of 0.1 m/s^2. The response of the structure results then in the amplitude of 0.15 m at the top level (corresponding to Z_0). The equations of motion are solved by using six modes in the expansion (20). The 4th-5th order Runge-Kutta algorithm numerical is used to solve the equations of motion numerically assuming the damping ratios $\zeta_1 = 0.003$ for the cable lateral mode, and $\zeta_2 = 0.3$ for the mass lateral mode and $\zeta_M = 0.3$ longitudinal mode, respectively. The response plots in figure 4 and figure 5 show the lateral displacements $\overline{v}_M$ of the mass, and the longitudinal motions u_M (coupled with the lateral motions) vs. time, respectively. The plots demonstrate resonance behavior of the system. The frequency spectra of the lateral response and the longitudinal response of the mass are shown in figure 6 and figure 7, respectively. The dominant frequency of the longitudinal response is twice the frequency of the excitation (tuned to the lateral resonance frequency of the mass). Thus, the adverse scenario could potentially arise when the longitudinal frequency of the mass near twice the lateral natural frequency of the mass. The longitudinal motions of the mass could then execute the lateral mode of the cable – mass system through the autoparametric coupling.

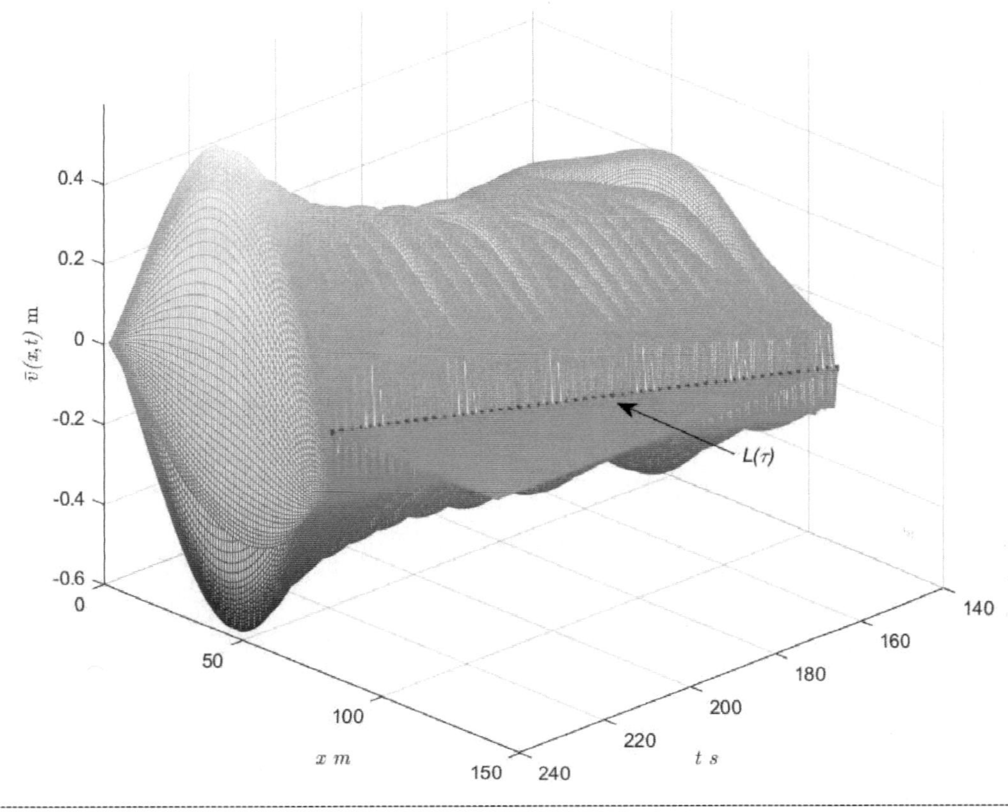

Figure 3. The lateral response $\overline{v}(x,t)$ of the cable - mass system.

4. Conclusions

The dynamic behaviour of a vertical cable – mass system moving slowly within a tall host structure subject to long period seismic excitation is considered in this paper. The proposed mathematical model accommodates the nonlinear effects of cable stretching and is used to determine the response of the system under the excitation caused by low frequency sway motions of the host structure. The case study presented in the paper demonstrates the effectivness of the proposed modelling approach to predict the dynamic behaviour of the system.

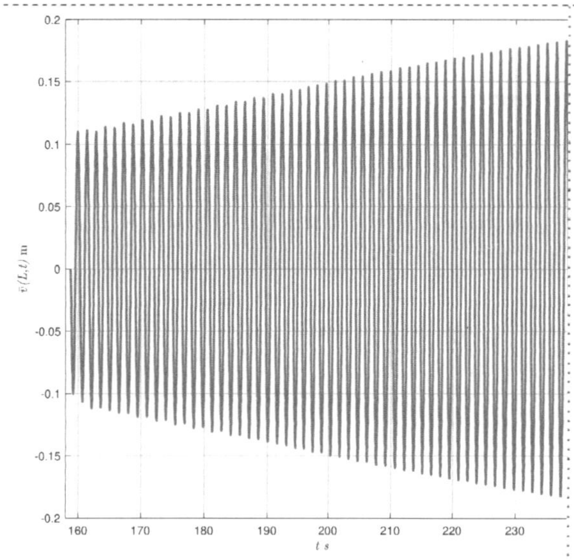

Figure 4. The lateral displacements $\overline{v}_M = \overline{v}(L,t)$ of the mass.

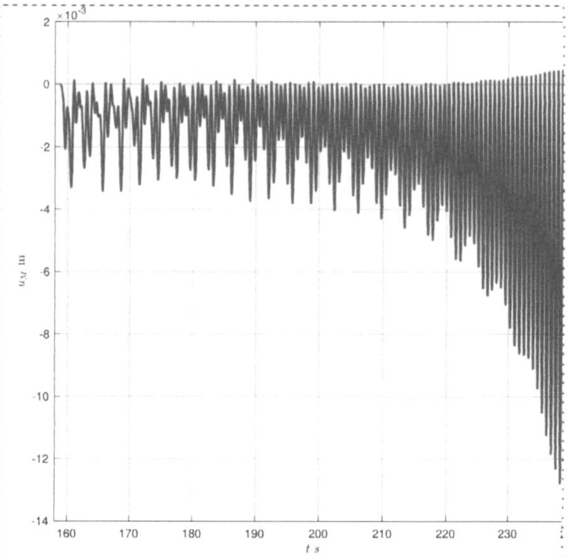

Figure 5. The longitudinal displacements $u_M(t)$ of the mass.

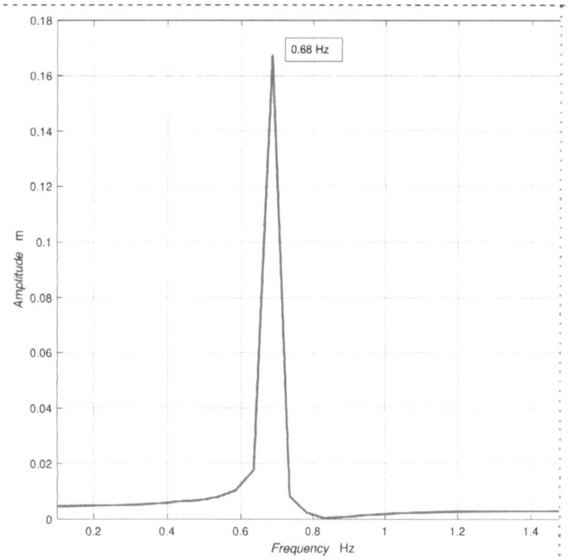

Figure 6. The frequency spectrum of the lateral response $\overline{v}_M = \overline{v}(L,t)$ of the mass.

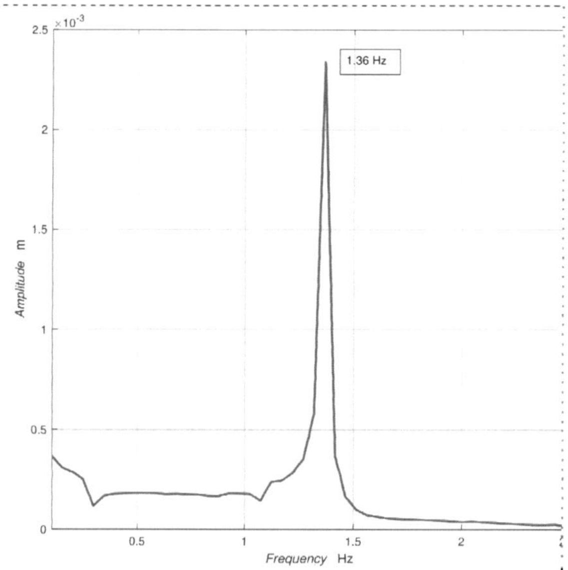

Figure 7. The frequency spectrum of the longitudinal response $u_M(t)$ of the mass.

References

[1] Hu R P, Xu Y L and Zhao X 2017 Long-Period Ground Motion Simulation and its Impact on Seismic Response of High-Rise Buildings *Journal of Earthquake Engineering* doi: 10.1080/13632469.2017.1286617

[2] Yao G C 2001 Performance of Passenger Elevators in Taiwan *Earthquake Engineering and Engineering Seismology* **3**(2) pp 17-26

IOP Conf. Series: Journal of Physics: Conf. Series **1048** (2018) 012005 doi:10.1088/1742-6596/1048/1/012005

[3] Saito T 2016 Response of High-Rise Buildings under Long Period Earthquake Ground Motions *International Journal of Structural and Civil Engineering Research* **5**(4) pp 308-314

[4] Meirovitch L 1990 *Dynamics and Control of Structures* (New York: John Wiley)

[5] Nayfeh A H and Pai P F 2004 *Linear and Nonlinear Structural Mechanics* (New York: John Wiley)

[6] Kaczmarczyk S and Iwankiewicz R 2017 Gaussian and non-Gaussian stochastic response of slender continua with time-varying length deployed in tall structures *International Journal of Mechanical Sciences* **134** pp 500-510

[7] Park S and Chung J 2014 Dynamic analysis of an axially moving finite-length beam with intermediate spring supports *Journal of Sound and Vibration* **333** pp 6742–6759

Elevator Rope Tension Analysis with Uneven Groove Wear of Sheave

Daisuke Nakazawa[1], Seiji Watanabe[1], Daiki Fukui[2], Ayaka Fujii[1] and Ken Miyakawa[1]

[1]Advanced Technology R&D Center, Mitsubishi Electric Corp., 8-1-1 Tsukaguchi Honmachi, Amagasaki, Hyogo, Japan
[2]Inazawa Works, Mitsubishi Electric Corp., Hishi-machi, Inazawa, Aichi, Japan

E-mail: Nakazawa.Daisuke@df.MitsubishiElectric.co.jp

Abstract. Traction elevators are suspended by multiple ropes. Uneven sheave wears affect the rope tension during the elevator operation. If the tension condition reaches to the critical traction ratio, a rope slip occurs. As the rope slip also affects the rope tension, a tension calculation with rope slip is necessary to evaluate the relation between the groove wear and the rope tension. In this paper, a tension evaluation model is derived by including the rope slip behavior and then the influence of the rope tension due to the groove wear is evaluated.

1. Introduction

In the traction elevator system, a car is suspended by multiple ropes. Ideally, each rope tension should be maintained as the balanced condition. However, the tension fluctuates due to the elongation of the rope, the variation of rope stiffness or the uneven sheave groove wear. If the tension of each rope is unbalanced severely, the elevator system is out of condition. Therefore, an analytical method is required to evaluate the rope tension behavior. Since the uneven groove wear of the sheave in high-rise elevators causes a significant tension change even in a small amount of wear, it is more important to establish the tension analysis with the groove wear than other factors. Therefore, clarification of the relationship among the groove wear, tension and traveling distance can contribute to the elevator's system design. Concerning the influence of the groove wear, the relationship between the amount of wear and the tension is investigated by Togawa et al [1]. While the above work mainly focuses on the experimental study of the phenomenon, an analysis of the tension change and the influence of groove wear have not been studied in detail. In this paper, firstly we describe the elevator's vertical vibration model with multiple ropes. Secondary, the mechanism of rope tension behavior with the uneven groove wear is explained. Finally, as the proposed model also includes the rope slip, the influence of the slip due to the unbalanced tension is evaluated by simulation and experiment.

2. Analytical Model

We derive an analytical model of the elevator that considers tension difference among the multiple ropes, and the rope slip on the sheave. Figure 1 shows the proposed model. In the model, ropes wrapped around a traction machine are connected to a car and a counterweight (CWT) via shackles. The ropes are modeled as springs. To evaluate the slip of the ropes on the sheave, the parts of ropes on the sheave are modeled as intensified masses. The equation of elevator's vertical motion of the elevator system is derived by the following equation.

$$\mathbf{M}\ddot{\mathbf{x}} + \mathbf{C}\dot{\mathbf{x}} + \mathbf{K}\mathbf{x} = \mathbf{F} \tag{1}$$

where **M**, **C** and **K** are the inertia matrix of the system, the damping matrix and the stiffness matrix respectively. **F** is the external force vector which contains the gravity force, the traction force and the torque by the traction machine. The elevator's transient motion is calculated by the numerical integration of equation (1).

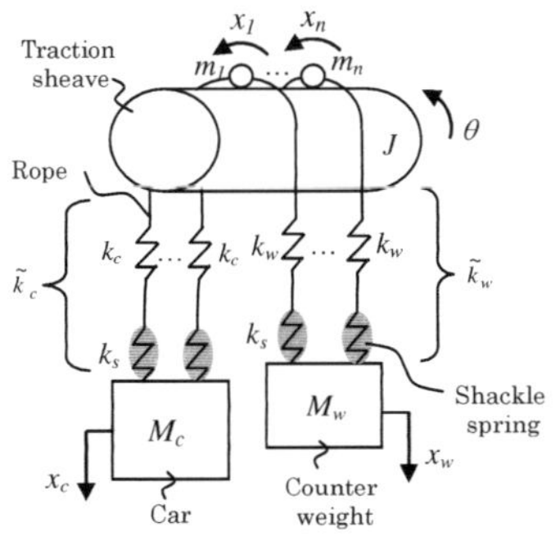

Figure1. Elevator model

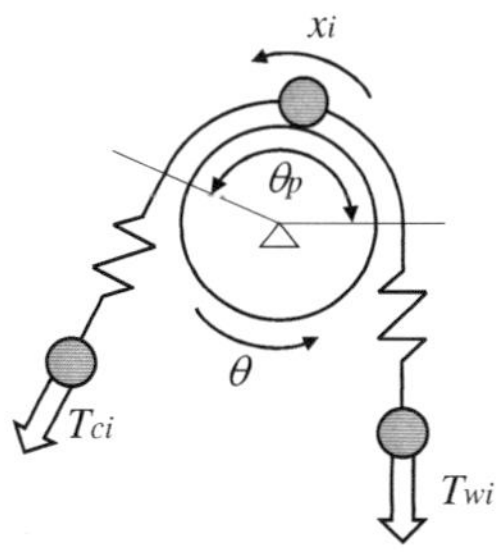

Figure2. Slip model

$$\mathbf{M} = diag\left[M_c \quad M_w \quad m \quad \cdots \quad m \quad J\right] \in \mathbf{R}^{(n+3)\times(n+3)}$$
(2)

$$\mathbf{K} = \begin{bmatrix} n\tilde{k}_c & 0 & -\tilde{k}_c & -\tilde{k}_c & \cdots & -\tilde{k}_c & 0 \\ 0 & n\tilde{k}_w & \tilde{k}_w & \tilde{k}_w & \cdots & \tilde{k}_w & 0 \\ -\tilde{k}_c & \tilde{k}_w & \tilde{k}_c + \tilde{k}_w & 0 & \cdots & 0 & 0 \\ -\tilde{k}_c & \tilde{k}_w & 0 & \tilde{k}_c + \tilde{k}_w & \cdots & 0 & 0 \\ & & & & \ddots & & \\ -\tilde{k}_c & \tilde{k}_w & 0 & 0 & \cdots & \tilde{k}_c + \tilde{k}_w & 0 \\ 0 & 0 & 0 & 0 & \cdots & 0 & 0 \end{bmatrix} \in \mathbf{R}^{(n+3)\times(n+3)} \quad (3)$$

$$\mathbf{F} = \left[M_c g \quad M_w g \quad f_{s1} \quad \cdots \quad f_{sn} \quad u - r\Sigma_{i=1}^{n} f_{si}\right]^T \in \mathbf{R}^{n+3} \quad (4)$$

$$\mathbf{x} = \left[x_c \quad x_w \quad x_1 \quad x_2 \quad \cdots \quad x_n \quad \theta\right]^T \in \mathbf{R}^{n+3} \quad (5)$$

Note that the damping matrix **C** consists of the damping elements which are proportional to the stiffness ones. Each element of **K** is quantified by the specification of the rope and the shackle spring stiffness. **C** is determined by the comparison with the experimental results which are shown in the following sections. The model parameters are shown in Table 1 where m is the intensified mass of the rope on the sheave. Other part mass of the rope is included in M_c or M_w.

Table1. Model parameters

M_c	Car mass	M_w	CWT mass
m	Rope mass on the sheave	x_c	Displacement of car
x_w	Displacement of CWT	x_i	Displacement of rope mass on the sheave
θ	Rotation angle of sheave	r	Sheave radius
J	Sheave moment of inertia	g	Gravity acceleration
$\tilde{k}_c$	Car side rope stiffness with shackle spring	$\tilde{k}_w$	CWT side rope stiffness with shackle spring
u	Torque	n	Number of the ropes

The rope slip can be checked by the ratio of the car side rope tension T_{ci} and the CWT side rope tension T_{wi}, as shown in figure 2. The ratio is compared with the critical traction ratio of the sheave Γ which is defined as

$$\Gamma = e^{\mu\kappa\theta_p} \quad (6)$$

where μ is the coefficient of friction, κ is the coefficient of the groove, θ_p is the wrapping angle of the rope on the sheave. In equation (4), f_{si} $(i=1,\ldots,n)$ is the force acting between the rope and the sheave. The force is switched according to the condition of the slip as follows.

(i) No slip condition

The condition that the rope does not slip is written as $T_{ci} < \Gamma T_{wi}$ where T_{ci} is larger than T_{wi}. On the other hand, if T_{ci} is smaller than T_{wi}, the no slip condition is given by $\Gamma T_{ci} > T_{wi}$ [2]. Therefore, by summarizing the equations, the no slip condition is

$$1/\Gamma < T_{wi}/T_{ci} < \Gamma \quad (7)$$

In this case, the rope and the sheave move with the same velocity, because the tension ratio of T_{ci} and T_{wi} is within the critical traction ratio. Hence, the following constraint equation is given as the no slip condition.

$$\bar{C}_i = r\dot{\theta} - \dot{x}_i = 0 \quad (i = 1, 2, \cdots n) \tag{8}$$

Each constraint equation can be summarized as the vector form

$$\bar{\mathbf{C}} = \left[\bar{C}_1, \cdots, \bar{C}_i, \cdots \bar{C}_n\right]^T = \mathbf{0} \tag{9}$$

Differentiating equation (9) with respect to time yields

$$\frac{\partial \bar{\mathbf{C}}}{\partial t} = \frac{\partial \bar{\mathbf{C}}}{\partial \dot{\mathbf{x}}}\frac{d\dot{\mathbf{x}}}{dt} = \bar{\mathbf{C}}_{\dot{\mathbf{x}}}\ddot{\mathbf{x}} = 0 \tag{10}$$

where $\bar{\mathbf{C}}_{\dot{\mathbf{x}}}$ is Jacobian of $\bar{\mathbf{C}}$ which can be written as

$$\bar{\mathbf{C}}_{\dot{\mathbf{x}}} \equiv \frac{\partial \bar{\mathbf{C}}}{\partial \dot{\mathbf{x}}} = \begin{bmatrix} 0 & 0 & -1 & 0 & 0 & \cdots & 0 & r \\ 0 & 0 & 0 & -1 & 0 & \cdots & 0 & r \\ 0 & 0 & 0 & 0 & -1 & \cdots & 0 & r \\ \vdots & \vdots & \vdots & \vdots & \vdots & \ddots & \vdots & \vdots \\ 0 & 0 & 0 & 0 & 0 & \cdots & -1 & r \end{bmatrix} \in \mathbf{R}^{n\times(n+3)} \tag{11}$$

Then, equation (1) and (10) can be written in the matrix equation as

$$\begin{bmatrix} \mathbf{M} & \bar{\mathbf{C}}_{\dot{\mathbf{x}}}^T \\ \bar{\mathbf{C}}_{\dot{\mathbf{x}}} & \mathbf{0} \end{bmatrix}\begin{bmatrix} \ddot{\mathbf{x}} \\ \mathbf{f}_s \end{bmatrix} = \begin{bmatrix} \bar{\mathbf{F}} - \mathbf{C}\dot{\mathbf{x}} - \mathbf{K}\mathbf{x} \\ \mathbf{0} \end{bmatrix} \tag{12}$$

$$\bar{\mathbf{F}} = \begin{bmatrix} M_c g & M_w g & 0 & \cdots & 0 & u \end{bmatrix}^T \in \mathbf{R}^{n+3} \quad , \mathbf{f}_s = \left[f_{s1}, \cdots, f_{si}, \cdots f_{sn}\right]^T \in \mathbf{R}^n$$

where constraint force $\mathbf{f}_s$ is calculated as the vector of Lagrange multipliers.

(ii) Slip condition
When the tension ratio of T_{ci} and T_{wi} is beyond the critical traction ratio, the rope experiences the slip on the sheave to keep the ratio equal to the critical traction ratio. Then the friction force f_{si} in equation (13) acts on the rope and the sheave, respectively.

$$f_{si} = \begin{cases} (\Gamma - 1)T_{wi}, & T_{wi} / T_{ci} < 1/\Gamma \\ (1 - \Gamma)T_{ci}, & T_{wi} / T_{ci} > \Gamma \end{cases} \tag{13}$$

3. Tension Analysis without Rope Slip

In this section, we show the tension change mechanism under the uneven groove wear condition by simulations and experiments. At first, we evaluate the tension behavior with respect to the car position without the rope slip. In the following discussion, we suppose that there is no passenger in the car, and the elevator consists of two ropes and the one of the sheave groove has a certain amount of wear. As the number of the ropes is two, the degree of freedom in the model equation (1) is five.

The groove wear is equivalent to the decrease of the sheave radius as shown in figure 3. It causes the difference of each rope's winding length on the sheave when the car goes up or down. This effect to the rope length causes the tension change as shown in figure 4. The horizontal axis shows the normalized car position. Its origin corresponds to the bottom floor, and the top floor is set to one. On the other hand, the vertical axis shows the tension which is normalized by the car side tension at the bottom floor. The mechanism of the tension change in the upward motion illustrated in figure 4 can be explained as follows;

1) The length of the rope 2 on the worn groove in the car side becomes longer than the rope 1 by the car upward motion. Also the length of the rope 2 in the CWT side becomes shorter than the rope 1, because the amount of winding on the sheave is smaller than one of the rope1.
2) The rope length difference causes the tension imbalance. The tension of rope 2 in the car side and the tension of rope1 in the CWT side decrease according to the car upward motion as shown in figure 4(a).
3) Since the rope length in the car side becomes shorter by the car upward motion, the tension changes sharply. On the other hand, as the rope length in the CWT side becomes longer, the tension change is

slower than the car side. In figure 4(b), the tension ratio of rope 2 becomes larger and it finally exceeds the critical traction ratio.

The tension change mechanism described above can be also expressed by the following functions of the car position.

$$T_{c1} = \frac{M_c g}{2} + K\frac{\beta x}{1-x+\delta},\quad T_{c2} = \frac{M_c g}{2} - K\frac{\beta x}{1-x+\delta},\quad T_{w1} = \frac{M_w g}{2} - K\frac{\beta x}{x+\delta},\quad T_{w2} = \frac{M_w g}{2} + K\frac{\beta x}{x+\delta} \tag{14}$$

$$K = EA/L \tag{15}$$

In the above equations, βx is the change of the rope length due to the groove wear where β is defined as $\beta = (1\text{-}r_2/r_1)/2$. $\delta = OH/L$ is the ratio of the remaining rope length OH on the top floor. x is the normalized car position which is defined as $x=(TR\text{-}x_c)/TR$ where TR is the elevator shaft height. E, A and L are the Young's modulus, the cross section area of the rope, and the distance from the bottom to the top floor, respectively. In Eq.11, the second term of the right hand side in all four equations represents the tension change as the curved line in figure 4(a). Note that we ignore the stiffness of the shackle spring in the derivation of equation (14) for the simplicity. When the traction ratio reaches the critical traction ratio, the rope2 on the worn groove slips with respect to the sheave in the actual system.

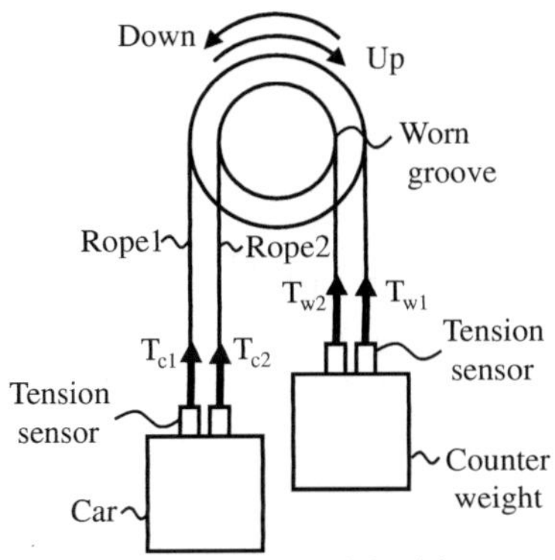

Figure3. 2-rope elevator model with groove wear

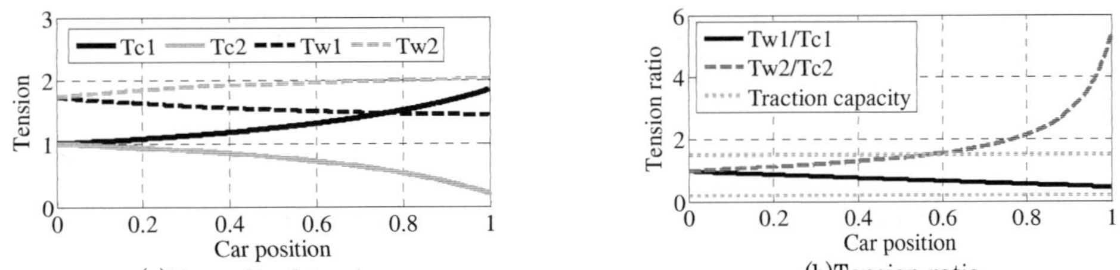

(a)Normalized tension (b)Tension ratio

Figure4. Simulation result of rope tension without slip (upward motion)

4. Tension Analysis with Rope Slip

In this section, the rope tension is measured by an elevator testbed to evaluate the rope tension behavior with rope slip. The configuration of the testbed is the same as figure 3. The tension sensors are attached to each rope end. Table.2 shows the conditions of the experiments. Figure 5 shows the comparison result between the simulation and experiment. As shown in figure 5, the simulation corresponds to the experimental data. The tension in the car upward motion behaves according to the following steps.

A) The tension change occurs based on 1) and 2) described in the previous section. The tension ratio of the rope2 reaches the critical traction ratio Γ. Then the rope 2 slips on the sheave.

B) Even after the tension ratio of the rope 2 reaches the critical traction ratio, the tension tries to change due to the length difference of the rope by the groove wear. As the traction ratio can't exceed the critical traction ratio, the tension ratio sticks to Γ, and the rope continues to slip until arriving at the top floor.

IOP Conf. Series: Journal of Physics: Conf. Series **1048** (2018) 012006 doi:10.1088/1742-6596/1048/1/012006

Table2. Experimental conditions

Elevator shaft height [m]	9.4
Car mass [kg]	292
CWT mass [kg]	457
Amount of wear [mm]	0.6
Number of ropes	2
Tension sensor (Load cell)	KYOWA LCW-C-10KN25SA3

When the tension ratio reaches the critical traction ratio, the slope of tension curve changes as shown in figure 5. Since the car and the CWT are suspended by multiple ropes, a tension change of a certain rope affects the tension balance. Thus, the tensions of the other ropes also change accordingly. Since the tension ratio with rope slip x_s can be written as $(T_{w2}-k_w x_s)/(T_{c2}+k_c x_s)=\Gamma$ where $k_w=K/(x+\delta)$ and $k_c=K(1-x+\delta)$, the amount of rope slip x_s can be expressed by the following equation.

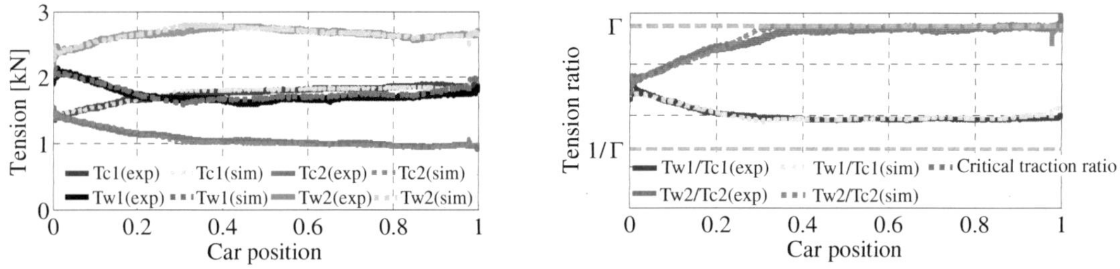

(a)Tension (b)Tension ratio

Figure5. Rope tension with rope slip behavior (Upward motion)
(Solid Line: Experiment, Dotted Line: Simulation)

$$x_s = \frac{T_{w2}-\Gamma T_{c2}}{D}, \quad D = K\left\{\frac{1}{x+\delta}+\Gamma\frac{1}{1-x+\delta}\right\} \tag{16}$$

Each of the rope tensions ($\hat{T}_{c1}$, $\hat{T}_{c2}$, $\hat{T}_{w1}$, $\hat{T}_{w2}$) after the slip occurrence is given by the following equations.

$$\hat{T}_{c1} = T_{c1} - K\frac{x_s}{1-x+\delta}, \quad \hat{T}_{c2} = T_{c2} + K\frac{x_s}{1-x+\delta}, \quad \hat{T}_{w1} = T_{w1} + K\frac{x_s}{x+\delta}, \quad \hat{T}_{w2} = T_{w2} - K\frac{x_s}{x+\delta} \tag{17}$$

The tension change is mitigated by the effect of the second term of each equation in equation (17), and that corresponds to the slope change of the tension in figure 5(a).

In the following part, we evaluate the relationship between the tension and the amount of the groove wear by the simulation. figure 6 shows the calculated tension during the car upward motion with various groove wear conditions. The vertical axis shows the tension which is normalized by the car side tension at the bottom floor. The slip starting position is shifted to the lower floor by a larger amount of groove wear as shown in figure 6(b). As a larger amount of wear increases the length difference of each rope, the tension ratio reaches Γ at the lower floor. Note that the tension follows the same line after the slip as shown in figure 6(a). To evaluate the above behavior, we transform the first equation of equation (17) by using equation (14) and equation (16). The tension after the rope slip can be written as the following equation.

$$\hat{T}_{c1} = \frac{M_c g}{2} - \left(\frac{M_w g}{2} - \Gamma\frac{M_c g}{2}\right)\frac{K}{D(1-x+\delta)} \tag{18}$$

Since equation (18) does not include β, it can conclude that the tension after the rope slip doesn't depend on the amount of wear, and the effect of the wear is eliminated by the rope slip.

IOP Conf. Series: Journal of Physics: Conf. Series **1048** (2018) 012006 doi:10.1088/1742-6596/1048/1/012006

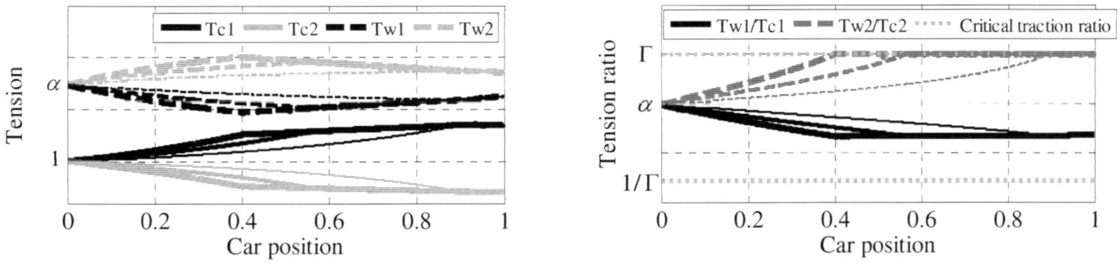

(a)Normalized tension (b)Tension ratio

Figure6. Simulation result about the influence of the groove wear

(Groove wear condition : Thick line × 1.5, Medium line × 1 ,Thin line × 0.5)

Finally, we discuss the rope tension behavior under the round trip condition. Figure 7 shows the rope tension with the rope slip during the downward motion of the return trip. When the car runs downward from the top floor, the car side rope tension T_{c2} increases and the CWT side rope tension T_{w2} decreases. This is because the amount of winding against the rope2 is smaller than the rope1. The rope1 also indicates the tension change with respect to the tension change of the rope2. Then the rope1 begins to slip on the sheave due to the above tension change. Figure 8 shows the tension behavior in the second and the third round trips in a row. The tension shows a hysteresis loop, and it follows the same line in every run. According to equation (18), the rope tension after the slip doesn't depend on β. Thus, the tension at the top or bottom floor is determined independently from the amount of wear. Therefore, if the initial tension condition is the same in every run, it concludes that the tension shows the same hysteresis loop.

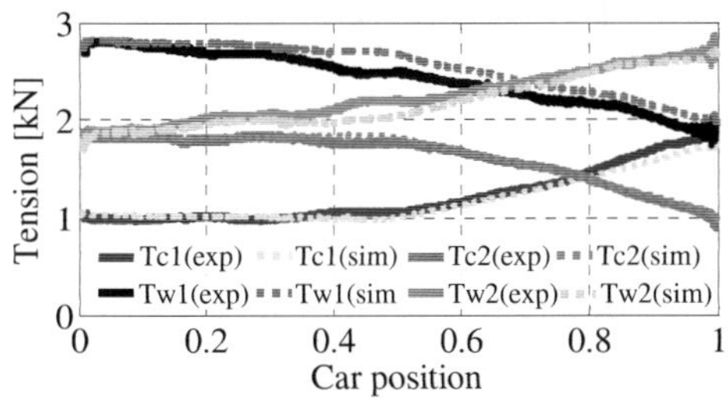

Figure7. Rope tension with rope slip
(Downward motion in first round trip)
(Solid Line: Experiment, Dotted Line: Simulation)

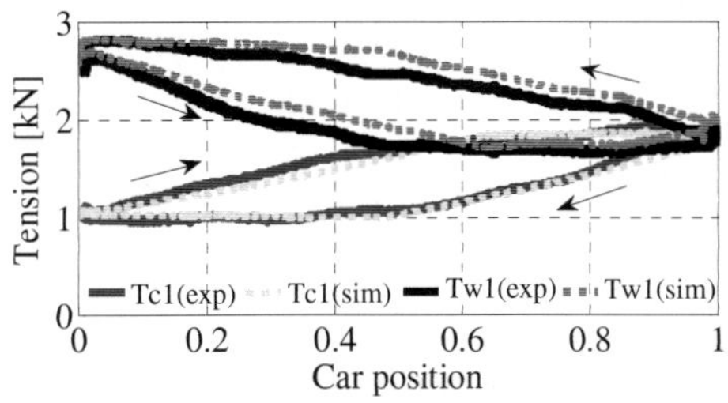

Figure8. Rope tension with rope slip
(consecutive round trips)
(Solid Line: Experiment, Dotted Line: Simulation)

5. Conclusions

In this study, the rope tension modeling of the elevator system was described. The proposed model can evaluate the rope slip behavior. The mechanism of the tension change due to the increase of the sheave groove wear was explained by the simulation and the experimental results. Under the condition that the car runs between the bottom floor and the top floor, we get the following results.

(a) The slip starting position is shifted to the lower floor.

(b) The rope on the worn groove slips when the car runs upward. On the other hand, the counterpart rope slips in the downward motion.

(c) The relationship between the rope tension and the car position shows the hysteresis loop, which is the same shape in every round trip.

Since the proposed model can properly simulate the tension behavior, the model can be applied to estimate the influence of the unbalanced tension conditions for the elevator system design.

References

[1] K. Togawa, S. Arai, M. Uwatoko: *Influence of Traction Sheave P.C.D. Difference on Sheave and Rope* (in Japanese), Proceedings of Elevator, Escalator and Amusement Rides Conference, Tokyo, Japan, 2012, pp.31-34.

[2] Gina Barney: Education & Training Sweet $e^{f\alpha}$?, ELEVATORI November/December, pp69, (Volpe Editore srl, Milan 2001)

The influence of shape on the vibratory response of variable beams

D Rodriguez[1a], G Rodriguez[2a] and I Herrera[3a]

[a]Escuela de Ingenierías Industriales, Universidad de Extremadura, Avda. de Elvas s/n, 06006 Badajoz, Spain

[1]daniel.tecno4@gmail.com, [2]gaspar.rodriguez.jimenez@gmail.com, [3]iherrera@unex.es

Abstract. Pure analytical solution for variable cantilever beams subjected to free bending oscillations and inertial forces is presented in this paper. Following the Euler-Bernoulli scheme for light slender cross section beams and Henri Bouasses's analytical procedure the in-plane linear motion differential equation is deduced. Taking into account the beam's shape a generalized non-dimensional differential equation is derived and characterized by only four geometric parameters. Beam's shape whose cross section area and moment of inertia variation obey a power law with respect to the longitudinal distance to the tip section of the bar and truncation are considered. The differential equation is analytically solved using the symbolic *Wolfram Mathematica* module for a number of typical shapes from uniform to zero area tip section with linear to cubic root variation and both rectangular and circular cross sections. The natural frequencies are determined and compared. Applications of results to high-rise building design and variable beams found in nature like tree stem are suggested.

1. Introduction

The main question that could be assessed is what shape of a variable beam is the best for a certain function or application considering its dynamic response. In the case of high-rise buildings both uniform cross section like beams and variable beams have been applied. A very important parameter of a building structure for seismic response is its natural frequency. It is of major importance to get a natural frequency as high as possible in order to avoid coupling or resonances. In the case of stem trees was found that pine, spruce and birch taper curve is clearly non-uniform non linear. What frequencies and mode shapes can be expected for these deformable solids? Let's see how the Euler-Bernoulli scheme for slender like beams could assess for that.

Although a numerical approach based, for instance, on the finite element method can be extremely useful to solve specific problems, it is also of great importance to get an analytical solution in order to predict which parameters are essential to the variable beams behavior. Pure theoretical analysis of beams are not only possible, but well developed as well since the nineteenth century. The dynamics of a beam, characterized by its uniform cross section, such as a rectangular cross section, is widely understood. Texts books like that of Belluzzi [1] reviews and pay attention to the dynamics of variable beam. However, truncated variable beams are still hard to be studied from a merely theoretical approach. Some studies have been made using different strategies, such as Bessel functions (Julián Muñoz González [2], 2009; De Rosa and Auciello [3], 1996) and orthogonal polynomials (Caruntu

[4], 2007) among others. Truncated beams with circular cross section were also approached in [3], while many authors approached the case of truncated rectangular beam. Several papers have been published solving the problem of a certain variation for the width, such as linear (Sanger [5], 1968), power variation (Wang [6], 1967) or stepped beams (Mazzei [7], 2014). Also Stanton and Mann ([8], 2010) analytically approached discontinuous Euler-Bernoulli beams, while different boundary conditions, such as rotational springs at both ends were taken into account by Maurizi and Bellés ([9], 1993).

The main purposes for this paper are reaching both an exact and yet easy solution to vibrations of variable beams using symbolic software interface. Although complicated frequency equations will be obtained for simple cases such as a beam with linearly variable thickness, this direct method will prove exact and yet easily applicable to a large variety of beam shapes and cross section shapes. The advantages of a pure theoretical non-dimensional solution make it possible to predict the particular solution for any case - including truncated beams – and study the response of different kinds of variation, such as linear or square root, on the natural frequencies and mode shapes.

2. Theoretical Procedure

According to the scheme in figure 1 the Euler-Bernoulli equation for vibrations of slender beam of ideal length L, truncated at section $x=v$, Young modulus $E(x)$, mass density ρ, variable cross area $S(x)$ and inertia $I(x)$ takes the form

$$\frac{\partial^2}{\partial x^2}\left[E(x)I(x)\frac{\partial^2 w}{\partial x^2}(x,t)\right] + \rho S(x)\frac{\partial^2 w}{\partial t^2}(x,t) = 0 \tag{1}$$

where w stands for the lateral displacement due to bending, t is the time and x the longitudinal coordinate - its origin being located at the virtual beam's tip section of zero area, pointing towards the embedment. If a uniform medium is considered, constant E modulus will apply. The cross area $S(x)$ and inertia $I(x)$, considering a power law variation, can be expressed by

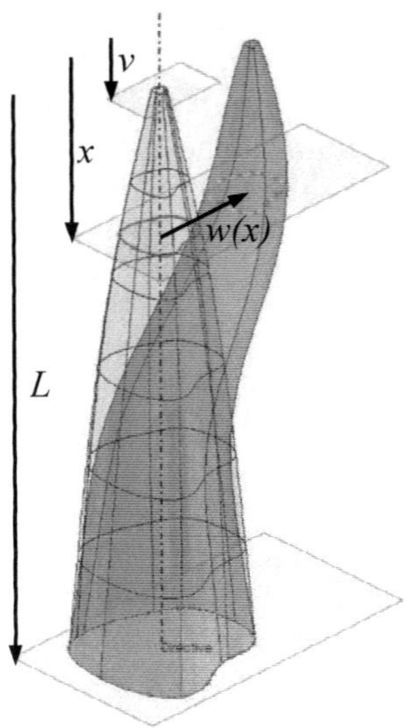

Figure 1. Generic variable beam

$$S(x) = sx^n \tag{2}$$
$$I(x) = ix^m \tag{3}$$

where s, i, n and m are determined to fit a particular cross section and beam's geometry - for example, for an ideal rectangular cross section ($v=0$) of linear thickness variation, constant width b and maximum thickness h: $n=1$, $s=bh/L$, $m=3$ and $i=bh^3/(12L^3)$, which do not depend on x nor t.

We state separated functions for x and t for the lateral displacement

$$w(x,t) = u(x)\sin(2\pi\omega t) \tag{4}$$

where u is a function that depends exclusively on x and ω is the frequency of the free oscillations. Substituting (2-4) into (1) for a beam of constant Young modulus E and assuming $\omega \neq 0$:

$$4\pi^2\omega^2\frac{\rho s}{Ei}u(x)x^n = \frac{d^2}{dx^2}\left[x^m\frac{d^2u(x)}{dx^2}\right] \tag{5}$$

Applying a generalized Bouasse's manipulation [10] a change of variable x to a new dimensionless variable $y=Kx$ can be performed for easiest integration and systematic solution. A like change of variables is performed in [11]. Then equation (5) reduces to:

$$u(y)y^n = \frac{d^2}{dy^2}\left[y^m\frac{d^2u(y)}{dy^2}\right] \tag{6}$$

If:

$$K = \left(4\pi^2\omega^2\frac{\rho s}{Ei}\right)^{[n-m+4]^{-1}} \tag{7}$$

K has always $[L]^{-1}$ dimension, regardless of n and m. From (2) and (3) it is easy to prove that s has $[L]^{2-n}$ dimension and i $[L]^{4-m}$ dimension.

Equation (6) represents a general differential equation valid for most slender variable beams whose geometry can be expressed as (2) and (3). This includes a wide variety of possible geometries and the cases we are interested in for smooth section variations which has been revealed to be a frequent case in nature [12]. See some geometries that apply in tables 1 and 2. A general case has to be approached directly by (5) without performing the above explained change of variable.

Now it is far more interesting to solve the problem following the simplified equation (6) and applying a non-dimensional analysis. Then, the result will be based on the dimensionless frequency ω^* given by:

$$\omega^* = \frac{\omega L}{\sqrt{E/\rho}} \tag{8}$$

and a new dimensionless K: $K^* = KL$. An explicit definition for ω^* can now be obtained substituting (8) in (7):

$$\omega^* = \left[\frac{i(K^*)^{n-m+4}}{4\pi^2 s L^{n-m+2}}\right]^{1/2} \tag{9}$$

Equation (6) is then solved using mathematical software symbolic interface -such as *Wolfram Mathematica*– . Figure 2 shows the display dump of the solution in which x represents the dimensionless y, *C[j]* for C_j, BesselI, BesselJ and MeijerG are Mathematica functions that were applied by the function *DSolve* [13] of the Wolfram Mathematica platform[14].

```
{{y → Function[{x}, ((BesselI[1, 2 √x] + BesselJ[1, 2 √x]) C[1]) / (2 √x) -
  (i (BesselI[1, 2 √x] - BesselJ[1, 2 √x]) C[3]) / (4 √x) + C[4] MeijerG[{{}, {}}, {{-1/2, 1/2}, {0, 0}}, x^2/16] +
  C[2] MeijerG[{{}, {}}, {{0, 0}, {-1/2, 1/2}}, x^2/16]]}}
```

Figure 2. Display dump for the Wolfram Mathematica symbolic solution.

The boundary conditions for the truncated cantilever beam must now be imposed: null displacement and rotation on the fixed section *u(K*)=0; u'(K*)=0;* and null bending moment and shear force on the free end section (defined by *x=v* or dimensionless *v*=v/L*) *u''(v*K*)=0; u'''(v*K*)=0*. This leads to a set of four equations with the four unknown integration constants due to (6) is a fourth order differential equation. Then, using the symbolic interface, the compatibility of the four equations is forced (the determinant of the matrix form for the set of equations for a vector with the unknowns to be zero) which leads to an algebraic equation dependent on *K** which is the frequency equation.

```
- 1/(8192 B^9 v^7) i (2 B^3 (B v (39 + 8 B v) BesselI[0, 2 √(B v)] - √(B v) (39 + 32 B v) BesselI[1, 2 √(B v)] +
```

Figure 3. Starting piece of the analytic symbolic frequency equation where B represents *K**.

This process is well known when applied to uniform beams, and is fully described, for instance, in [15]. The exact form of the mode shapes can also be obtained. This can be described by the position of the different nodes.

As detailed above, as soon as a beam's geometry can be described by Eq. 2 and Eq. 3, regardless of the section shape - rectangular, circular, etc. -, a complete study on frequencies and modes can be approached from (6). Naturally it is of high importance to be able to fully describe the beam's geometry, for only this way can the variables *s* and *i* be obtained as functions of geometrical parameters such as the diameter at the embedment section, the thickness, etc.

As an example, some cases are solved for different geometries - therefore different values for *n*, *m*, and also *s* and *i* -. All of them are considered tip-ended (*v = 0*).

Let us consider a rectangular cross section cantilevered beam of constant depth *b* whose thickness h_x increases by the power law $h_x = h(x/L)^a$ - *a* is its exponent- to get the maximum thickness *h* in the fixed end. Then we get from (2) and (3):

$$s = \frac{bh}{L^a};\ i = \frac{bh^3}{12L^{3a}}; n = a; m = 3a \tag{10}$$

Substituting upon (9):

$$\omega^* = \frac{(K^*)^{2-a}}{4\pi\sqrt{3}} \cdot \frac{h}{L} \tag{11}$$

Equation (11) shows how the dimensionless frequency ω^* depends linearly on the slenderness ratio *h/L* of the beam. For *h/L=0*, $\omega^*=\omega=0$.

Let us consider another important variable beam which is that of circular cross section cantilevered beam whose diameter d_x increases by the power law $d_x = d(x/L)^a$ to get the maximum diameter d in the fixed end where a is the exponent. Now we get from (2) and (3):

$$s = \frac{\pi d^2}{4L^{2a}};\ i = \frac{\pi d^4}{64L^{4a}};n = 2a; m = 4a \tag{12}$$

Substituting upon (9):

$$\omega^* = \frac{(K^*)^{2-a}}{8\pi}\cdot\frac{d}{L} \tag{13}$$

Equation (13) shows how the dimensionless frequency ω^* depends again linearly on the slenderness ratio d/L of the beam. Equations (11) and (13) can be synthesized and simplified setting the slope C with respect to the slenderness ratio, named frequency coefficient and defined by:

$$\begin{aligned} &\omega^* = C\cdot\frac{h}{L} \text{ for rectangular cross section where } C = \frac{(K^*)^{2-a}}{4\pi\sqrt{3}} \\ &\omega^* = C\cdot\frac{d}{L} \text{ for circular cross section where } C = \frac{(K^*)^{2-a}}{8\pi} \end{aligned} \tag{14}$$

The same process can be followed with different cases. As long as equations (11) and (13) are valid, the problem reduces to determine the corresponding solution values K^*_j $(j=1,2,3,...)$ from the frequency equation, that is, its roots. These are correspondent to the natural frequency ω_1 $(j=1)$ and its harmonics ω_j $(j=2,3,...)$ of the beam.

3. Results

Tables 1 and 2 show the first three frequency coefficients C_1, C_2 and C_3 and the volume ratio for the volume of the variable beam to that for the equivalent uniform beam V_v/V_u for a number of variable beams, plus the uniform rectangular and circular cases.

Table 1. First three natural frequency coefficients and volume ratio of variable rectangular-section cantilever beams.

Beam type	*Drawing*	*a*	*n*	*m*	V_v/V_u	C_1	C_2	C_3
Rectangular, uniform	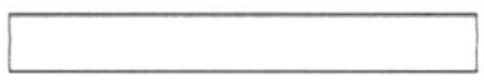	0	0	0	1	**0.1615**	**1.0124**	**2.8346**
Rectangular, cubic root variation		1/3	1/3	3	3/4	0.1963	0.9279	2.3148
Rectangular, square root	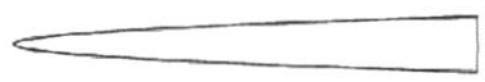	1/2	1/2	3/2	2/3	0.2113	0.8774	2.0678
Rectangular, 1.8 root		1/1.8	1/1.8	3/1.8	0.6429	0.2159	0.8595	1.9874
Rectangular, 1.6 root		1/1.6	1/1.6	3/1.6	0.6154	0.2214	0.8363	1.8882
Rectangular, 1.4 root		1/1.4	1/1.4	3/1.4	0.5833	0.2279	0.8055	1.7630
Rectangular, 1.2 root	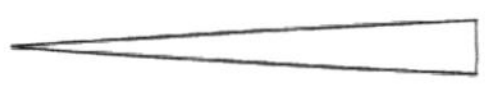	1/1.2	1/1.2	3/1.2	0.5455	0.2356	0.7624	1.5999
Rectangular, linear beam		1	1	3	0.5	**0.2442**	0.6987	1.3792

Table 2. First three natural frequency coefficients and volume ratio of variable circular-section cantilever beams.

Beam type	*Drawing*	*a*	*n*	*m*	V_v/V_u	C_1	C_2	C_3
Circular, uniform	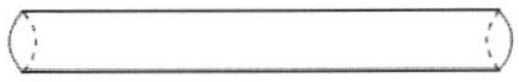	0	0	0	1	**0.1399**	**0.8767**	**2.4549**
Circular, cubic root variation	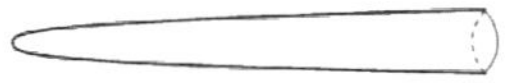	1/3	2/3	4/3	0.6	0.2146	0.9059	2.1607
Circular, square root	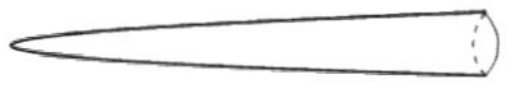	1/2	1	2	1/2	0.2506	0.9040	2.008
Circular, 1.8 root	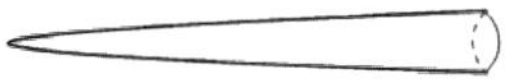	1/1.8	2/1.8	4/1.8	0.4737	0.2623	0.9011	1.9565
Circular, 1.6 root		1/1.6	2/1.6	4/1.6	0.4445	0.2766	0.8961	1.8914
Circular, 1.4 root		1/1.4	2/1.4	4/1.4	0.4118	0.2945	0.8872	1.8069
Circular, 1.2 root		1/1.2	2/1.2	4/1.2	0.3750	0.3174	0.8713	1.6927
Circular, linear beam		1	2	4	1/3	**0.3469**	0.8414	1.5300

It is then possible to obtain the exact values - according to Euler-Bernoulli theory - for every ω^*_j only by multiplying the corresponding C_j - table 1 or 2 - by the slender ratio of the sample *h/L* or *d/L* according to (14). For instance, a rectangular linearly variable beam, with *h/L*=0.1 gives the results shown in table 3 for the non-dimensional frequencies ω^*_i. Variable rectangular beam frequencies f_i can also be expressed in terms of f_{ui}, the *i-th* frequency for the rectangular uniform beam. It can be derived from (8) that the ratio f_i/f_{ui} -also shown in Table 3- is coincident with the ratio C_i/C_{ui} where C_{ui} is the slope *C* for the *i-th* frequency rectangular uniform beam -provided that both beams are the same length and material-. Results shown in table 1 and 2 have been validated using the approximate method of Rayleigh-Ritz and FEM *SolidWorks* simulation tool [16] and those shown in bold by comparison with specific exact solutions in the literature [1, 10, 15].

Table 3. First three natural frequencies of a rectangular, linear beam

Beam type	*h/L*	ω^*_1	ω^*_2	ω^*_3	f_1/f_{u1}	f_2/f_{u2}	f_3/f_{u3}
Rectangular, linear beam	0.1	0.0244	0.0699	0.1379	1.5117	0.6902	0.4866

Now, once the frequencies are known, their associated mode shapes can be calculated by substituting the value of any of the frequencies ω^*_j, in the set of boundary conditions equations, the four constants appearing on the solution for (6) are solved, and, unmaking the change of variables, the full solution for every mode shape is obtained. As an example, a linear rectangular beam vibrating in the horizontal plane can be analyzed and compared with results given in [10] (chapter II, page 123). Bouasse obtains *0.3290* for the node of the second harmonic - as position from the origin divided by length - and *0.2136* and *0.5495* for both nodes corresponding to the third harmonic. Plotting the full solution, as shown in figure 4, for each mode following the steps above, *0.3287* is obtained as the node's position for the second mode, while *0.2133* and *0.5490* stand for the third mode.

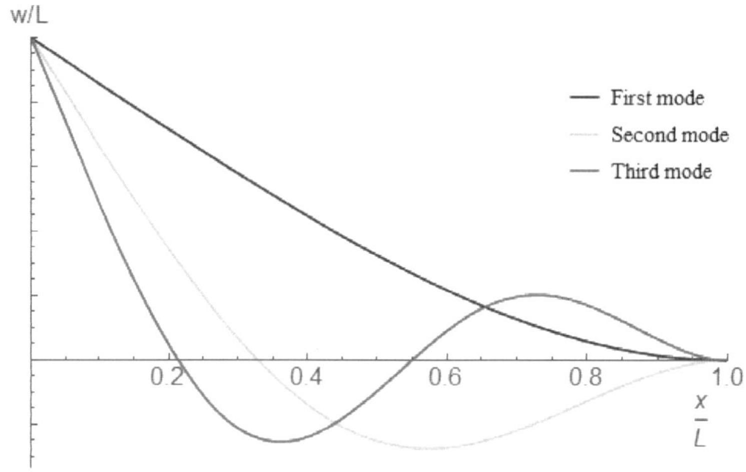

Figure 4. Natural modes for a linear rectangular beam vibrating in the horizontal plane, cantilevered at $x/L = 1$.

4. Conclusions

A non-dimensional generalized differential equation and boundary conditions were derived from classical Euler-Bernoulli equation for free vibrations of variable-shape cantilever beams by applying a generalized Bouasse change of variable with certain limitations for the cross section area and inertia power laws. It was found that the fundamental frequencies for tip ended rectangular and circular cross section variable beams (14) are linearly dependent on the slenderness ratio (h/L or d/L) whose slope C (14) depends on the thickness or diameter power law exponent a. The equations are solved using the Wolfram Mathematic symbolic interface. Exact frequencies for rectangular/circular cross section of a varying thickness/ diameter cantilever beams are determined for a number of different power laws. It was found that the fundamental frequency of a linear, square root or cubic root shaped rectangular/circular beam is significantly higher than that of a similar uniform beam although its volume is lower than that of the uniform one. The second and third frequencies are lower. Finally, by viewing (5) for the rectangular cross section values shown in (10), it can be easily concluded that the depth b -constant along the beam's span- has no influence on the natural frequencies.

References

[1] Belluzzi, O 1976 *Ciencia de la Construcción*. Vol 4, First Edition. Aguilar.
[2] Muñoz, J 2012 *Vibraciones de Vigas de Sección Variable*. First edition. Editorial Aerodinámica Española,
[3] De Rosa M A, Auciello N M 1996 Free vibrations of tapered beams with flexible ends. *Computers and Structures* **60** (2), 197–202.
[4] Caruntu D I 2007 Classical Jacobi polynomials, closed-form solutions for transverse vibrations. *Journal of Sound and Vibration* **306** (3–5), 467–494.
[5] Sanger D J 1968. Transverse vibration of a class of nonuniform beams. *Journal of Mechanical Engineering Science* **16**, 111–120.
[6] Wang H C 1967 Generalized hypergeometric function solutions on the transverse vibrations of a class of nonuniform beams. *Journal of Applied Mechanics*. **34E**, 702–708.
[7] Mazzei A and Scott R 2014 Vibrations of discretely layered structures using a continuous variation model. *Allemang R (ed) Topics in modal analysis II,* vol **8**. Springer International Publishing, pp 385–396.
[8] Stanton S C and Mann BP 2010 On the dynamic response of beams with multiple geometric or material discontinuities. *Mechanical Systems and Signal Processing* **24**: 1409–1419.

[9] Maurizi M J and Belle´s P M 1993 Natural frequencies of onespan beams with stepwise variable cross section. *Journal of Sound and Vibration* **168**: 184–188.
[10] Bouasse H 1927 *Verges et plaques*, cloches et carillons, Paris, Delagrave.
[11] Kim Y C 2010 An asymptotic solution and the Green's function for the transverse vibration of beams with variable properties pp. 34-38, ISSN 1225-0767.
[12] Laasasenaho J 1982 Taper curve and volume functions for pine, spruce and birch, Communicationes Instituti Forestalis Fenniae. The Finnish Forest Research Institute, Helsinki.
[13] Stephen Wolfram 2003 *The Mathematica Book*. Fifth Edition. Wolfram Media.
[14] Wolfram Mathematica Software release 10.4
[15] Rao S 2011 *Vibraciones Mecánicas*. Fifth edition. Pearson.
[16] Rodríguez D 2016 Estudio dinámico de barras a partir del modelo de Euler-Bernoulli. Aplicación a álabe de aerogenerador. TFG, University of Extremadura.

IOP Conf. Series: Journal of Physics: Conf. Series **1048** (2018) 012008 doi:10.1088/1742-6596/1048/1/012008

Effects of Muscle Length and Physiological Cross Sectional Area on Muscle Force Production: a Comparative Study

F Romero-Sánchez[1], F J Alonso[1], J Barrios-Muriel[1] and G Rodríguez-Jiménez[2]

[1]Department of Mechanical Engineering, Energy and Materials, University of Extremadura. Avda. de Elvas s/n, 06006 Badajoz, Spain
[2]Department of Mechanical Engineering, University of Seville, Camino de los Descubrimientos s/n. 41092 Seville, Spain.

[1]fromsan@unex.es, [1]fjas@unex.es, [1]jorgebarrios@unex.es

Abstract. The estimation of muscular forces is useful in several areas such as biomedical or rehabilitation engineering. The analysis of the forces that produce a given movement (inverse dynamics, ID) or the movement induced by a set of muscle forces or activations (forward dynamics, FD) are typical problems that need the description of muscle mechanical properties. Moreover, to solve the indeterminacy problem in the ID analysis, optimization schemes are required. Several optimization methods (static optimization, dynamic optimization, augmented static optimization) and optimization criteria (minimum metabolical cost of transport, minimum sum of muscle stresses, time-integral cost of activations, torque-tracking) are available in the literature to that end. In all the aforementioned problems muscle properties, such as muscle length or physiological cross sectional area (PCSA), play a key role in the development of consistent models to perform specific tests. In this work, a comparative study of the effects on muscle force production in muscles with different ratios between length and PCSA is presented. Its relevance in the traditional problems faced in biomechanics is also studied.

1. Introduction

The analysis of muscle force activity is an essential exercise to understand how the force is generated to perform a movement. Movement Biomechanics deals with the analysis of the forces that produce a given movement (the inverse dynamics problem, ID) and the analysis of the movement produced by a given set of muscle forces (forward dynamic problem, FD). Both problems need to be described by muscle models. The work of Hill [1] stablished the basis for a parametric model description. Since then, different models have been described in the literature (see [2,3] for traditional models or [4,5] more complex models based on fractional order dynamics). Such models are described mainly by the muscle length, muscle contraction velocity and activation level. By using these models, once the current state is known (muscle length and contraction velocity), it is possible to obtain the output, i.e., the muscle force for different sets of activations that combined provide the net joint torque in a joint and, therefore, the movement. On the contrary, to the aforementioned description of the FD problem, in the ID problem the inputs are the spatial coordinates of the body and the ground reaction forces. Both allow to obtain the net joint torques and joint reaction forces. The muscle forces are then obtained by distributing the net joint torque between the muscles spanning the joint. To do so, different optimization schemes can be used [6-8] where the usual cost function depends on the muscle stress, which is directly related to the

IOP Conf. Series: Journal of Physics: Conf. Series **1048** (2018) 012008 doi:10.1088/1742-6596/1048/1/012008

physiological cross-sectional area (PCSA). As it is shown, traditional problems in Biomechanics depends on the muscle macroscopic structure, and therefore, either the force production in the FD problem or the muscle force distribution in the ID are conditioned by it. For the muscles shown in figure 1, for the same level of activation the force production will be different in a slender muscle compared to a wider muscle as well as the muscle stress. In this work, we study the differences in the resolution of both typical problems in Biomechanics, for simple examples regarding muscles of the lower limbs were the biggest differences between slender muscles can be found.

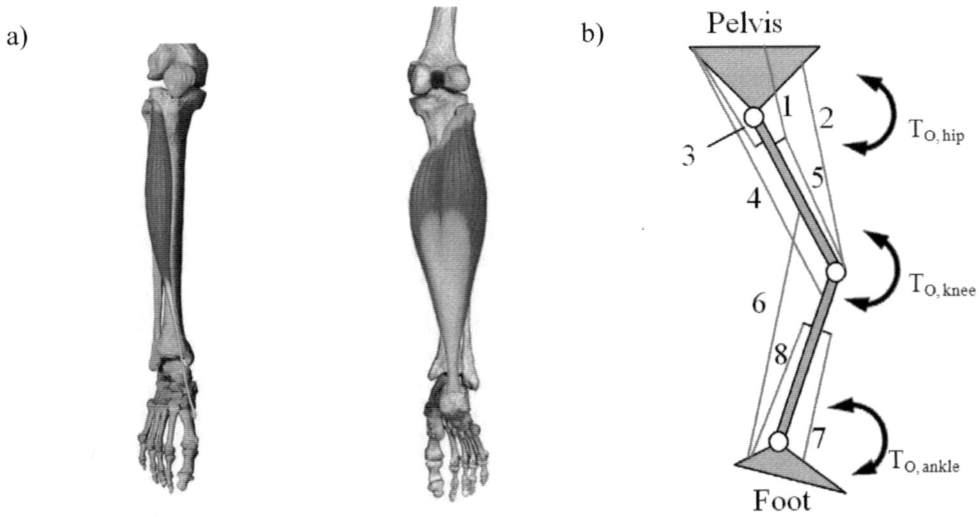

Figure 1. a). Differences in muscle length and physiological cross sectional area on shank muscles. Left: Anterior view of the shank: *tibialis anterior* muscle. Right: Posterior view of the shank: *gastroccnemious*. b) Muscle groups used in this work: 1 – Iliopsoas, 2 – Rectus Femoris, 3 – Glutei, 4 – Hamstrings, 5 – Vasti, 6 – Gastrocnemius, 7 – Tibialis Anterior, 8 – Soleus. T_o represents the given torque at each joint.

2. Methods

In order to study the influence of different values of muscle length and physiological cross-sectional in muscle force production, we first present the problems in which these terms are involved, and then the methodology to test the influence of variations in those parameters.

2.1. The FD problem. Description of the muscle model.

In a previous work, the authors tested the different Hill-type dynamics formulations for muscle force estimation [9] that accounted for muscle parameters such as muscle length, muscle contraction velocity or activation level among others. In this work we use the description provided by Thelen [10], as the output is implicitly related to muscle parameters.

The contractile element in a Hill-type muscle model is responsible of the active force generation in muscle tissue. The force developed in this element (contraction dynamics) depends on two relationships, the force-length and the force-velocity relationships:

$$F^M = F_0^M \cdot a \cdot f_L(\tilde{l}^M) \cdot f_V(\tilde{v}^M) \tag{1}$$

where F_0^M is the maximum isometric force of the muscle, a is the activation level ($a \in [0,1]$), and $f_L(\tilde{l}^M)$ and $f_V(\tilde{v}^M)$ are the force-length and the force-velocity relationships, respectively. On the one hand, the force length relationship is defined as:

$$f_L(\tilde{l}^M) = \begin{cases} 0 \quad for \quad \tilde{l}^M \leq 1 \\ e^{-(\tilde{l}^M-1)^2/\gamma} \quad for \quad \tilde{l}^M > 1 \end{cases} \tag{2}$$

where $\tilde{l}^M = l^M / L_{opt}^{CE}$ is the normalized muscle length by the length in which the maximum force is obtained and γ a shape factor set to 0.45 in this work.

On the other hand, the force-velocity relationship is written as:

$$f_V(\tilde{v}^M) = \begin{cases} 0 \quad for \quad \tilde{v}^M \leq -1 \\ \dfrac{1+\tilde{v}^M}{1-\tilde{v}^M/K_{CE1}} \quad for \quad -1 < \tilde{v}^M \leq 0 \\ \dfrac{1+\tilde{v}^M f_v^{max}/K_{CE2}}{1+\tilde{v}^M/K_{CE2}} \quad for \quad \tilde{v}^M > 0 \end{cases} \tag{3}$$

where $\tilde{v}^M = v^M/(L_{opt}^{CE}/\tau_c)$ is the normalized muscle contraction velocity by a term that represents the optimal contraction velocity (τ_c is a time constant set to 0.1 in this work), and K_{CE1} and K_{CE2} are force-velocity shape factors (0.25 and 0.06 respectively in this work). For further details of this description see [10].

2.2. The ID problem. Description of the optimization scheme.

Since several muscles serve each joint of the skeletal system, muscle forces cannot be directly computed from joint moments. Several optimization methods (static optimization, dynamic optimization, augmented static optimization, large-scale static optimization) and optimization criteria (minimum metabolical cost of transport, minimum sum of muscle stresses, minimum hyper-extension of the joints, time-integral cost of activations, torque-tracking) are available in the literature. In this work we use an optimization scheme proposed by the authors in which the physiological information of the muscle is considered. The scheme is as follows:

1.- In the first step, the length and velocity of each muscle are obtained from generalized coordinates of the multibody model. The maximum muscle force histories F_{max}^M compatible with contraction dynamics are calculated supposing that the muscle activation are maxima at every instant $\mathbf{A}_m = [a_1, \ldots, a_N]^T = [1, \ldots, 1]^T$, i.e., the hypothetical maximum muscle force that correspond to the current state:

$$F_{max}^M = F_0^M \cdot 1 \cdot f_L(\tilde{l}^M) \cdot f_V(\tilde{v}^M)$$

2.- For each muscle, the muscle activation levels are obtained by solving the following optimization scheme:

$$\begin{aligned} min \quad J &= \sum_{j=1}^{n} \frac{a_j \cdot F_{max,j}^M}{C_j} \\ s.t. \quad & \mathbf{R}(\mathbf{A} \cdot \mathbf{F}^{\mathrm{M}}) = \mathbf{T} \end{aligned} \tag{4}$$

where $C_j = F_0^M/PCSA$ is the muscle stress, **T** is the net joint torque, $\mathbf{A} \cdot \mathbf{F}^{\mathrm{M}} = \left[a_1 \cdot F_{max,1}^M, a_2 \cdot F_{max,2}^M, \dots, a_n \cdot F_{max,n}^M\right]^T$ are the muscle forces scaled by the design variables and **R** is the constant matrix of equivalent moment arms of the different muscles. Once solved the optimization scheme, the F_{max}^M term can be scaled by the activations obtained in the optimization problem.

2.3. Experimental setup.

In order to test the influence of muscle length and PCSA, we defined two different experiments (figure 2a). Regarding the first experiment, to quantify the influence of muscle length, we solved the FD problem. As this problem requires the values of muscle length (l^M), muscle contraction velocity (v^M) and activation (a_j) levels to be performed, we recorded the gait from a healthy voluntary subject and then we performed an inverse kinematics analysis to obtain the muscle length and muscle contraction velocity. This process can be found in detail in, e.g., [11,12]. As stated before, there is a major number of slender muscles in the lower limbs that can be used for the proposed study. As activations cannot be measured *in-vivo* non-invasively, we used the expression proposed by [13] that relates the activation level with the electromyography (EMG) signal that can be measured from surface muscles. This relationship can be written as:

$$a_j(t) = \frac{e^{A \cdot u_j(t)} - 1}{e^A - 1} \tag{5}$$

where $a_j(t)$ is the activation level, $u_j(t)$ is the processed EMG signal and A is a non linear shape factor constrained between $-5 < A < 0$ [13] (in this work A = -4). By using the measured parameters. i.e., $l^M$and v^M from the ID of the registered motion and EMG signals on equations (2), (3) and (5), respectively, it is possible to obtain an estimate of the muscle forces in the FD problem.

The EMG signal was recorded in the tibialis anterior muscle by following the standard recommendations given by the SENIAM [14] using the Trigno™ Wireless System from Delsys, at a sample frequency of 2000Hz. The EMG signal was imported to Matlab (The Mathworks, Inc.) where the signal processing was performed. The EMG signal was rectified and a moving average (MOVAG) was applied to smooth the signal. Then, we used the recorded and processed EMG signals on (5) to obtain the activation levels. Then, by using equations (1)-(3) we obtained the muscle forces based on the muscle state, retrieved from the coordinates of the recorded motion. The motion was recorded by means of 12 Optitrack V100-R2 IR cameras that register the position of reflective markers placed according figure 2b. In order to solve the dynamic problem, we also recorded the ground reaction forces from a force plate Bertec FP4060-10-2000. In order to compare signals, and considering the gait as a relatively slow movement to reduce the influence of muscle contraction velocity, we used the same activation level as input for all the muscles and we studied the differences between lengths. Certainly, the obtained forces do not correspond to the ones expected during gait, nevertheless the results can be compared in terms of similar contraction velocity and activation levels, being muscle length the variable term.

Regarding the second experiment, in order to quantify the influence of PCSA, we solved the distribution of a given set of joint torques in the lower limbs between a number of muscles (figure 1b). To do so, we used the expressions presented in section 2.2. As the torque distribution provides the muscle force for the muscles acting at each joint based on the muscle stress, the results are directly comparable. Nevertheless, we solved the same problem for different values of muscle stress in order to also compare variations for the same muscle (this could be comparable to study the differences in muscle force between marathon runners and sprinters that develop the same conditions during motion, except for the value of PCSA). The data set for net joint torques was retrieved from [15], as it has been widely used in the literature.

a)

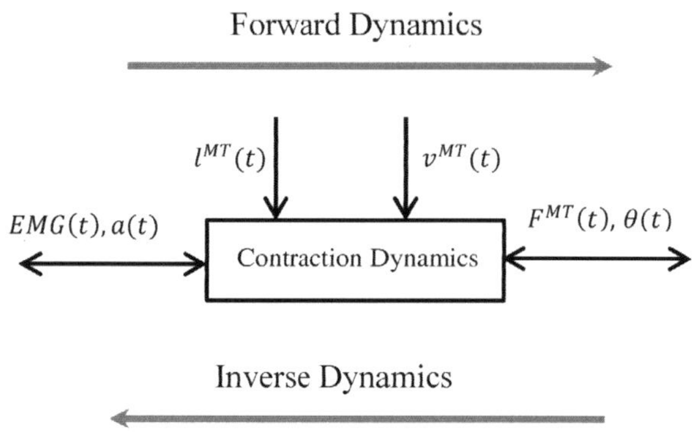

b)

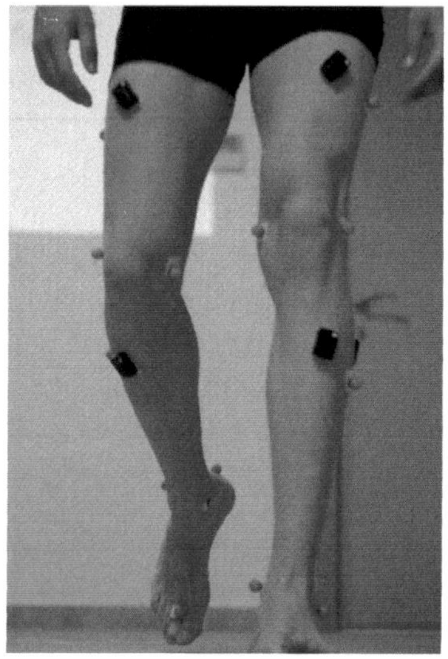

Figure 2. a) EMG electrodes and reflective markers placement. b) Schematic representation of the two problems faced at muscle level.

3. Results and discussion

Regarding the results of the FD analysis (figure 3) a clear relationship between muscle force and muscle length cannot be stated. For example, a slenderer muscle such as the tibialis anterior produces less force than a wider muscle, as, i.e., the gastrocnemius. This may be due to the tridimensional structure of muscle tissue that has not been considered in this study. Furthermore, the viscoelastic properties of the tendon, that complete the muscle structure, were not considered as they do not contribute to the force production. Nevertheless, its inclusion may modify the results in terms of analysis as they may modify the slenderness of the muscle (see figure 1, the gastrocnemius has a large portion of tendon that may modify the relationship between length and PCSA).

On the contrary, for the results of the ID problem (figure 3), it is shown a clear relationship between muscle force and PCSA. The wider the muscle is, the larger force is obtained. For a smaller PCSA the muscle stress is lower, and therefore the expected force production. A wider muscle supports larger stress. Results below 100% represent not only a thinner muscle, but also a proportion of the muscle that has been activated, that is directly related with the recruitment process of muscle fibres to exert a certain amount of force. Results over 100% represent an increment in the cross-sectional area of the muscle tissue.

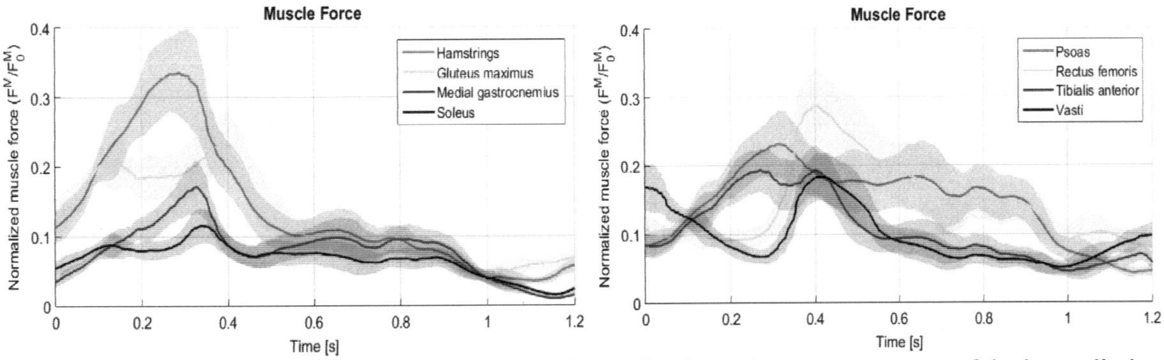

Figure 3. Forward dynamic solution. Muscle forces for the main actuator groups of the lower limbs. Top: red line: hamstrings; green line: gluteus maximus; blue line: medial gastrocnemius; black line: soleus. Bottom: red line: psoas; green line: rectus femoris; blue line: tibialis anterior; black line: vasti.

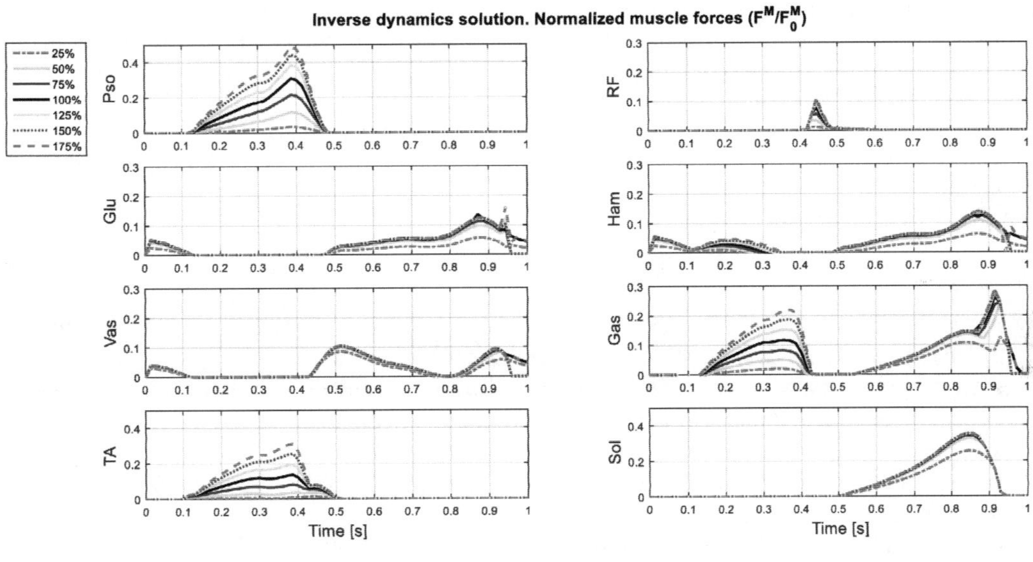

Figure 4. Inverse dynamic solution. Muscle forces for the main actuator groups of the lower limbs for different values of the PCSA. Top: red dashed line: 25% PCSA; green line: 50% PCSA; blue line: 75% PCSA; black line: 100% PCSA, cyan line: 125% PCSA; violet dotted line: 150% PCSA; orange dashed line: 175% PCSA.

Table 1. Fundamental parameters of the muscle [11, 12, 15]

L_{opt}^{CE} (cm)	PCSA (cm^2)	F_0^M (N)	$L_{opt}^{CE}/PCSA$ (cm^{-1})

Psoas	10.2	7.9	821	1,291
RF	8.1	13.9	663	0,582
Glu	20	30.4	1705	0,657
Ham	10.4	4.9	1770	2,122
Vas	9.3	16.8	7403	0,553
Gas	5.5	21.4	1639	0,257
TA	8.2	11.0	1528	0,745
Sol	5.5	58.8	3883	0,093

4. Conclusions

In this work the influence between muscle length and muscle section in force production for the traditional problems in Biomechanics has been presented. Although this is a preliminary work, the study must be completed by including the 3D structure of muscle tissue and the effects of the tendon and the surrounding viscoelastic tissue. Furthermore, a simple non-linear relationship between EMG signal and muscle activation has been used to overcome the problematic of the activation level (a(t)). The way in which muscle fibers are recruited it is not studied here, nevertheless it must be considered in future studies as it may influence the results in terms of PCSA.

References

[1] Hill A 1938 *Proc. R. Soc. Lond. B* **126** 843 136-95.
[2] Hatze H, 1977*Biol. Cybern.* **25** 103–19.
[3] Winters J M and Stark, L 1987 *Biol. Cybern.* **55** 403–20.
[4] Sommacal L, Melchior P, Cabelguen J-M, Oustaloup A, and Ijspeert A 2007 *Advances in fractional calculus* Springer 271–85
[5] HosseinNia S, Romero F, Tejado I, Vinagre B, and Alonso J 2012 *System Identification, International Federation of Automatic Control* **16** 1743–48.
[6] Ackermann M: *Dynamics and energetics of walking with prostheses*, PhD thesis, University of Stuttgart, Stuttgart, Germany, 2007.
[7] Ackermann, M and Schiehlen,W 2006 *Arch. Appl. Mech.*, **75**, 569–94.
[8] Alonso J, Romero F, Pàmies-Vilà R, Lugrís U, and Font-Llagunes J 2012 *Multibody Syst. Dyn.* **28**, 109–24.
[9] Romero F, and Alonso F J 2016 *Mechanical Sciences* **7** 1 19-29.
[10] Thelen D 2003 *J. Biomech. Eng.- T. ASME* **125** 70-7 2003.
[11] Yamaguchi G 2001 *Dynamic modeling of musculoskeletal motion: a vectorized approach for biomechanical analysis in three dimensions* Kluwer Academic Publishers Norwell MA.
[12] Zatsiorsky V and Prilutsky B 2012 *Biomechanics of skeletal muscles, Human Kinetics, Human Kinetics Publisher* Champaign USA.
[13] Sartori M, Reggiani M, Mezzato C, and Pagello E 2009 *IEEE International Conference on Advanced Robotics, ICAR 2009* Munich Germany 1–7.
[14] Hermens H J., et al. European recommendations for surface electromyography 1999 *Roessingh research and development* **8** 2 13-54.
[15] Winter D A 2009 *Biomechanics and motor control of human movement* John Wiley & Sons Waterloo Canada.

IOP Conf. Series: Journal of Physics: Conf. Series **1048** (2018) 012009 doi:10.1088/1742-6596/1048/1/012009

Key Innovations and Technologies in Vertical Transportation Systems since 1980

Rory S. Smith

University of Northampton, Northampton, UK

Rory.smith@northampton.ac.uk

Abstract. Vertical transportation systems have been rapidly evolving since the 1990's. Lifts without machine rooms have become common, many buildings of 100 or more floors have been built or are planned and new technologies for lifts are being introduced.

The common types of lifts produced in 1990 are explained as a baseline for explaining the innovations that have revolutionized lifts since then. These innovations include Variable Voltage Variable Frequency (VVVF) technology, the Insulated Gate Bipolar Transistors (IGBT), Permanent Magnet Alternate Current Motors (PMAC), Independent Wire Rope Core (IWRC) ropes, and Machine Room Less (MRL) lift technologies.

Two novel suspension means, coated steel belts and coated carbon belts, are new technologies that are changing lift designs are reviewed.

Systems that employ more than one lift per hoistway are explained. These systems can use ropes or they can be rope-less and employ magnetic levitation.

Two interesting new technologies that are profoundly changing the lift industry, Machine Learning and Robots are also examined.

1. The History of lifts

Lifts have existed for thousands of years. However prior to 1853 they had one major problem; if the suspension ropes failed the lift would fall and the passengers would be injured or die.

In 1853, Elisha Otis demonstrated a device, now known as the safety gear that arrested the lift's fall when the ropes were cut [1]. This device made lifts safe and the lift industry was started.

Advances in structural engineering made tall buildings possible and safe lifts made them practical.

In the approximately 140 years following the invention of the safety gear, lifts slowly evolved. In 1990 there were two main types of lifts these were Traction and Hydraulic.

2. Traction lifts

Traction lifts consist of a car, hoist ropes, a machine, and a counterweight. See Figure 1, Traction Components.

The ropes were made of steel strands wound around a hemp rope core.

The machine was a pulley connected to a motor. For high speed lifts, the motor was a Direct Current motor and was powered by physically large variable voltage drive. Slow speed lifts with speeds below 2.5 m/s used a special alternating current motor with small inefficient variable speed drive. The motor of the slower lifts was fitted with a reducing gear.

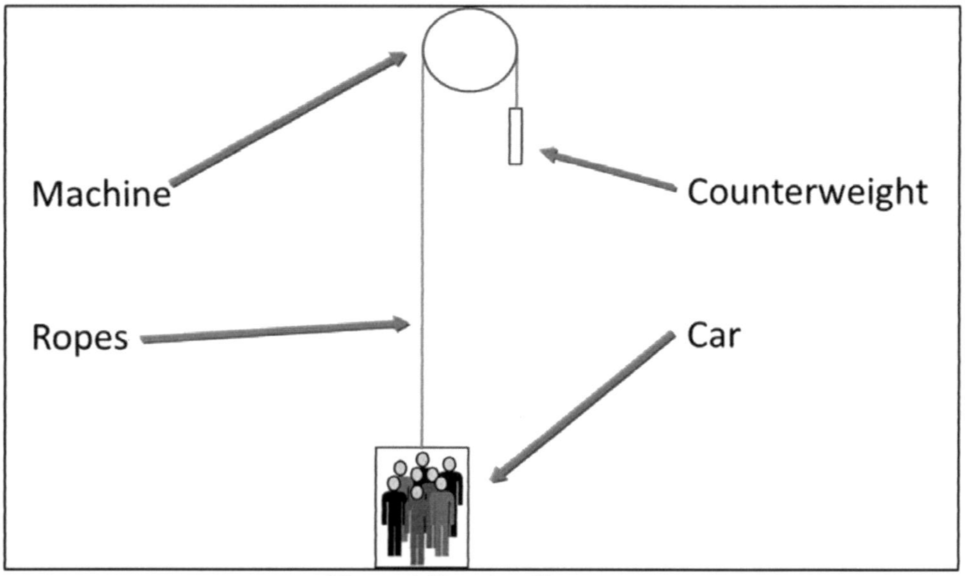

Figure 1,Traction Components

In very tall buildings, a variation of the traction lift called a double deck lift was sometimes utilized. In double deck lifts the car had two cabins stacked one on top of the other. Double deck lifts still had one very large machine and one very large counterweight.

3. Hydraulic lifts

For low rise buildings up to around 8 to 10 floors, hydraulic lifts were often used. Hydraulic lifts consist of an oil reservoir, an oil pump, a set of control valves and a hydraulic ram (piston and cylinder). To move the lift in the up direction oil was pumped from the reservoir to the hydraulic ram. To go down, a valve was opened allowing the oil to flow from the ram to the reservoir. See Figure 2, Hydraulic Components.

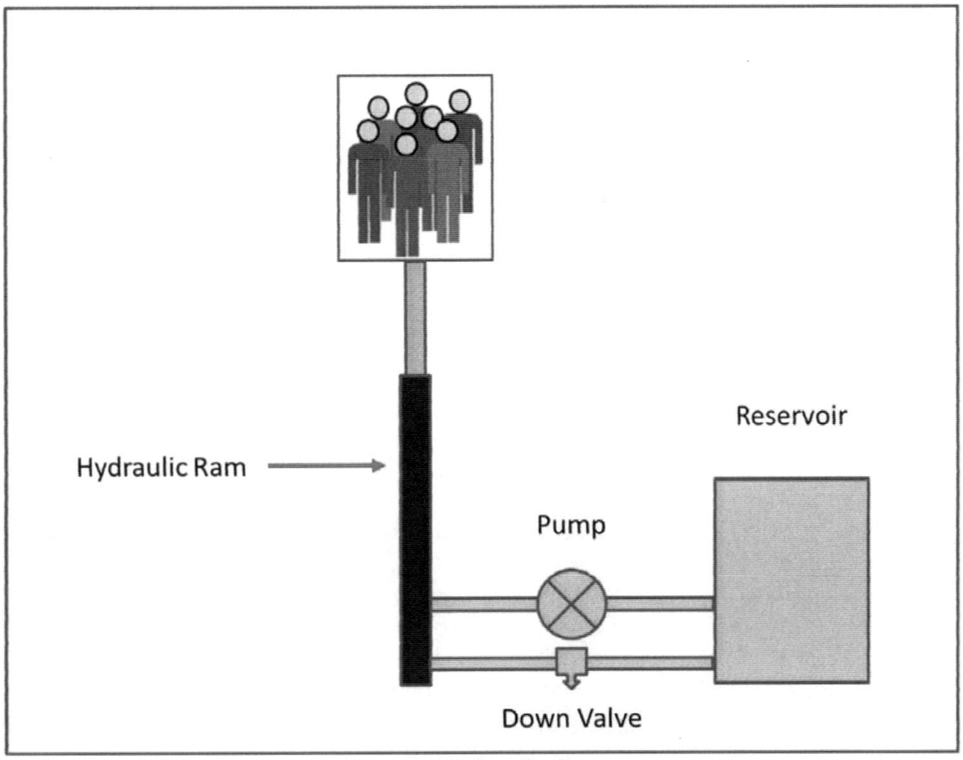

Figure 2, Hydraulic Components

Hydraulic elevators are lower in cost than traction elevators.

4. Key innovations that changed lift design

4.1 Variable Voltage Variable Frequency

The speed of an Alternating Current (AC) motor is directly proportional to the frequency of the power supply. In most of the world the AC power supplied is 50Hz. If a motor designed for 50Hz could be supplied with 25Hz it would rotate at half its design speed.

Lifts using VVVF technology were introduced in the 1980's. In order to create the variable frequency a technology known as Pulse Width Modulation (PWM) was employed [2] [3]. The switching transistors used for these applications were limited to a switching frequency of around 3,000 Hz. The sine waves produced would be made up of 3,000 pulses each second.

The 3,000 Hz made audible noise and caused vibration in the motor. To address these issues, large air core inductors were used. The inductors caused the VVVF drive to be very large. However, this new technology worked well.

4.2 IGBT

A new type of transistor was developed in the 1980's and found its way into lifts in the early 1990's. This type of transistor is known as the Insulated Gate Bi-Polar Transistor (IGBT) [4].

The IGBT could switch at frequencies above 10,000Hz. At these frequencies no inductors were required for vibration or noise control.

The VVVF variable speed drive using IGBT's was smaller than the Direct Current drives or the early VVVF drives.

4.3 IWRC

As previously mentioned, lift ropes were typically made up of steel strands wound around a non-load carrying core. In the 1990's a new type of rope was introduced known by its construction as Independent Wire Rope Core (IWRC). In this rope, the fibre core was replaced with a steel load carrying core [5].

An IWRC rope could carry more load than a traditional rope of the same diameter or ropes of smaller diameter can carry the same load. Using smaller diameter ropes has several significant advantages:

1. Pulleys must be 40 times the diameter of the rope. The smaller pulleys make the machine more compact.
2. Smaller pulleys must turn more revolutions per minute (RPM) for the same lift speed.
3. Higher rpm's means the motor must develop less torque for the same amount of power. (See formula 1 below) [6].
4. Lower torque motors are less expensive and physically smaller.

$$P = \frac{\tau \times 2\pi \times v}{60} \quad 1.$$

Where: P represents power
τ represents torque
v represents RPM

4.4 Permanent Magnet Alternating Current Motors

Lifts utilizing VVVF technology initially used alternating current induction motors. The induction motor had been in existence for around 100 years. In the late 1990's Permanent Magnet Alternating Current Motors (PMAC) started to be used for lifts.

PMAC motors are, for any given power and torque, smaller and more efficient than induction motors [7].

4.5 Combined technologies

By utilizing VVVF motor control using IGBT transistors along with PMAC motors and IWRC ropes it was now possible to make a very compact elevator.

Traditionally, lifts had an equipment room on the roof of the building where the machine and control were located. The area of this machine room was often larger than the area of the lift shaft.

In the late 1990's KONE elevator introduced a novel lift system that combined all these technologies and was so compact that no machinery space was required above the roof of the building [8]. The concept was extremely popular. KONE gained market share and all the other lift companies scrambled to develop similar products. The new lift type became known as an MRL, an acronym for Machine Room Less [8].

IOP Conf. Series: Journal of Physics: Conf. Series **1048** (2018) 012009 doi:10.1088/1742-6596/1048/1/012009

5. New Technologies

5.1 Belts

Two belt technologies were introduced.

In the early 2000's Otis elevator introduced a coated steel belt [9]. The belt has the steel strands of a round rope. However they are laid out horizontally and coated with polyurethane. This permits the use of a sheave with a very small diameter (100mm). Consequently, the motor torque is greatly reduced.

In 2012 KONE introduced a carbon fibre belt [10]. The principal advantage of the rope is its low weight. In very tall buildings, steel rope begins to be unable to support its own weight. The carbon fibre rope solves this problem.

5.2 Building Efficiency

One way to measure building efficiency is determine the net useable floor space of a building as a percentage of the gross area. If one visits a 100 story building, one will find that a very large percentage of the lower floors are dedicated to lift lobbies and shafts.

Double deck lifts are one of the methods that can be used to improve efficiency. The big drawback of double deck lifts is that the two cabins must travel together as they have one hoist machine and one counterweight. In the early 2000's ThyssenKrupp Elevator introduced a solution known as TWIN [11].

The TWIN lift consists of two complete lifts, two cabins, two machines, two sets of ropes, and two counterweights with the cars traveling on the same rails and using most of the same entrances. The individual cars can move independently.

An additional product was just recently introduced by ThyssenKrupp Elevator known as MULTI. In this scheme multiple lift cars can be in the same hoistway.

The lift cars a powered by linear motors and do not have hoist ropes [12]. The system uses one lift shaft for cars traveling up and another shaft for cars traveling down. At the top and bottom floors, the cars can transit horizontally from up to down shafts or vice versa.

A traffic analysis of this system performed by the author demonstrated a significant increase in net useable floor space when compared to a conventional lift system.

5.3 Robots

Professional service robots are appearing in the marketplace that will need to ride in lifts with humans [13].

There are four building types where professional service robots are being applied in increasing numbers. These building types are hotels, office buildings, residential buildings, and hospitals [14].

In hotels, service robots are being used for room service deliveries and for the delivery and movement of housekeeping carts.

In office, buildings robots are delivering packages and mail from sources outside the building as well as interoffice correspondence.

The growth of e-commerce has caused an increase of package deliveries to multi-story residential buildings. Robots are being used to make the final delivery from the lobby to the residential unit.

Hospitals have used pneumatic tube systems for delivery of medicines and the transport of medical records. Pneumatic tube manufacturers are now offering robots as an alternative to tubes [15].

Lift companies are developing interfaces to allow robots to call lifts and have the lifts take them to their destinations [16].

5.4 Machine Learning
Machine learning, a form of Artificial Learning (AI) is being applied to lifts in order to increase the up-time of lifts [17]. Machine learning makes predictions based on known properties learned from training data.

In a typical lift installation, a data collection device is connected to the lift that monitors the operations of the lift. Additionally any error codes issued by the lift's control system are also collected. The lift continuously sends this data to a cloud computing facility where it is analysed.

The machine learning system then schedules routine maintenance, defines maintenance tasks, and even schedules special maintenance visits to make repairs prior to a part failing.

Machine Learning will totally change the lift industry. Lifts will become more reliable and operating costs will be reduced.

6. Conclusion
It can be seen that since 1990, lifts have incorporated many technological changes. These changes have made lifts more efficient and more reliable.

The rate of change is increasing and Machine Learning and other forms of AI will further improve safety, efficiency, and reliability.

7. References

[1] *Elisha Otis* Available from: https://en.wikipedia.org/wiki/Elisha_Otis Accessed: 24 March, 2018

[2] *Pulse Width Modulation* Available from: https://www.techopedia.com/definition/9034/pulse-width-modulation-pwm Accessed: 23 March 2018

[3] *Variable Frequency Drive* Available from: https://en.wikipedia.org/wiki/Variable-frequency_drive Accessed: 23 March, 2018

[4] *Insulated Gate Bi-polar transistor* Available from: https://en.wikipedia.org/wiki/Insulated-gate_bipolar_transistor Accessed: 20 March, 2018

[5] *Traction Ropes for Traction Sheave Elevators* Available from: https://www.pfeifer.info/en/products-services/wire-ropes-in-application/mechanical-engineering/elevator-construction/elevator-operation/traction-ropes-for-traction-sheave-elevators/traction-ropes-for-high-rise-traction-sheave-elevators/ Accessed: 24 march, 2018

[6] *Torque* Available from: https://en.wikipedia.org/wiki/Torque Accessed: 19 March, 2018

[7] *AC induction Motors vs. Permanent Magnet Synchronous Motors* Available from: http://empoweringpumps.com/ac-induction-motors-versus-permanent-magnet-synchronous-motors-fuji/ Accessed: 23 March, 2018

[8] *The History, benefits and Concerns of the new MRL elevator Systems* Available from: https://www.buildings.com/article-details/articleid/3076/title/machine-room-less-mrl-elevators Accessed: 22 march, 2018

[9] Andrew J and Kaczmarczyk S 2011 *Systems Engineering of Elevators* Mobile: Elevator World

[10] *Ultra Rope* Available from: http://www.kone.com/en/products-and-services/solutions-for-special-buildings-and-large-projects/ Accessed: 21 March, 2018

[11] Smith R *TWIN A Revolution in Elevator Construction.* Institution of Electrical Engineers, Seminar on Intelligent Environments, University of Essex, UK. June 2005

[12] ThyssenKrupp AG *Linear motor for an Elevator* EP17708489 4 March, 2016 Authors: Smith R, et al

[13] *Executive Summary World Robotics 2017 Service Robots* Available from: https://ifr.org/downloads/press/Executive_Summary_WR_Service_Robots_2017_1.pdf Accessed 09 February, 2018

[14] Smith R 2018 *Robots, a New Class of Passenger* Symposium on Lift and Escalator Technologies Hong Kong

[15] Discussions with Bob Bauer, Savioke Robotics, December, 2017.

[16] Discussions with Phil Herget, Savioke Robotics, January, 2018.

[17] Smith R 2015 *The Internet of Things, Big Data, Machine Learning, and the Lift and Escalator Industry* Symposium on Lift & Escalator Technologies Northampton UK

IOP Conf. Series: Journal of Physics: Conf. Series **1048** (2018) 012010 doi:10.1088/1742-6596/1048/1/012010

The modal characteristics of high aspect ratio sailplane wings including the effects of bending and torsional rigidities

H Su[1,3] and J R Banerjee[2]

[1] Department of Engineering, University of Northampton, UK,
[2] Department of Mechanical Engineering & Aeronautics, School of Mathematics, Computer Science and Engineering, City, University of London, UK.
[3] Correspondence email address: huijuan.su@northampton.ac.uk

Abstract. The modal behaviour of two high aspect ratio sailplane wings with cantilever boundary conditions is investigated by applying the dynamic stiffness method and using the Wittrick-Williams algorithm as solution technique. The wing is idealised as an assembly of bending-torsional coupled beams for which the frequency dependent dynamic stiffness matrix is established. A Fortran computer program is developed to obtain the natural frequencies and mode shapes of the wings. The bending and torsional rigidities of each wing are varied and their subsequent effects on the natural frequencies and mode shapes are examined. A detailed parametric study with the variations of bending and torsional rigidities shows some interesting results which can be of practical help to the industry when designing such wings. The illustrative examples chosen for the baseline wings prior to changing their rigidities are representative of existing sailplanes. The results are presented and discussed and the paper concludes with some remarks.

1. Introduction

Sailplane wings are slender and flexible due to their high aspect ratios resulting from large spans and relatively short chords. As a consequence, they are easily prone to vibration problems. Modal analysis of sailplane wings is thus very important and plays an important role in their design. Analysis of this kind is an obligatory airworthiness requirement which is rigorously enforced by the civil aviation authorities. The purpose of this paper is to carry out such an analysis to predict the modal behaviour of sailplane wings by applying the dynamic stiffness method (DSM) and subsequently examining the effects of the wing rigidity properties on the modal characteristics. In order to provide the context of this research, interested readers may find references [1-6] useful as background studies.

One of the main motivations for modal analysis of aircraft wings originates from the fact that it is a fundamental prerequisite when carrying out the response analysis, particularly through the use of the normal mode method. It is well known that the finite element method (FEM) is generally used to solve such problems. Using the FEM, the stiffness and mass properties of all individual elements are assembled to form the overall stiffness and mass matrices of the structure. The typical eigenvalue problem is solved to provide the natural frequencies and the corresponding mode shapes of the structure upon imposing the boundary conditions. The FEM is numerically intensive and the degrees of freedom identified by the order of stiffness and mass matrices decide the number of eigenvalues, which are the natural frequencies. The higher order natural frequencies will be considerably less accurate. There is a powerful alternative to the FEM called the DSM for modal analysis of aircraft wings or other structures.

The DSM is based on an exact single dynamic stiffness element containing both the mass and stiffness properties of the element as the basic building block. The assembly procedure is the same as in the FEM, but a single dynamic stiffness element matrix is used for each structural component instead of separate mass and stiffness matrices to form the overall frequency-dependent dynamic stiffness matrix of the complete structure. The formulation in the DSM leads to a transcendental/nonlinear eigenvalue problem whereas a linear eigenvalue problem is generally encountered in the FEM. The best available solution technique to extract the eigenvalues in the DSM is to apply the Wittrick-Williams algorithm [7], which has received extensive coverage in the literature. The algorithm, which uses the Sturm sequence property of the dynamic stiffness matrix, is robust and ensures that no natural frequency of the structure is missed.

The paper focuses on the modal behaviour of two cantilever sailplane wings. The wings are idealised as an assembly of bending-torsional coupled beams [8-9]. The frequency dependent dynamic stiffness matrix of a bending-torsion coupled beam required for the analysis is outlined here. A computer program in Fortran is developed to obtain the natural frequencies and mode shapes of such wings using the Wittrick-Williams algorithm. The bending and torsional rigidities of each wing are then varied and their subsequent effects on the natural frequencies and mode shapes are investigated. A detailed parametric study with the variations of bending and torsional rigidities shows some interesting results which can be of practical help in the design of aircraft wings. The illustrative examples chosen for the baseline wings are representative of those of existing sailplane wings and the paper concludes with some remarks.

2. Theory

2.1. Formulation of dynamic stiffness matrix of a bending-torsional coupled beam

Figure 1 shows a bending-torsion coupled beam with length *L*, distance between the mass and elastic axes x_α, the bending and torsional rigidities *EI* and *GJ* respectively, the mass per unit length *m*, the polar mass moment of inertia per length about the *Y*-axis I_α. The bending displacement and torsional rotation are *h* and ψ. The coupled bending and torsional motions occur due to non-coincident mass and elastic axes which are respectively the loci of the centroid and shear centres of the beam cross-section. For an aircraft wing, it is not generally possible to realise a torsion-free bending displacement or a bending-free torsional rotation during its dynamic motion unless the load or the torque is applied through or about the shear centre. Given the perspective, a high aspect ratio non-uniform sailplane wing can be accordingly modelled as an assemblage of bending-torsion coupled beams. The representation in figure 1 is particularly relevant to analyse a high aspect ratio aircraft wing.

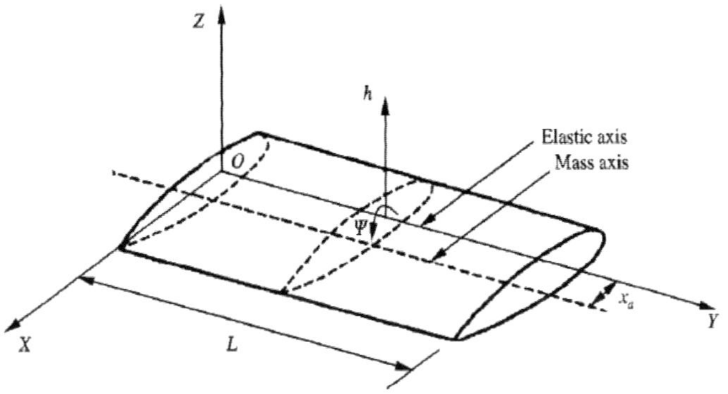

Figure 1. A wing idealised as a bending-torsion coupled beam.

The dynamic stiffness matrix of a uniform bending-torsion coupled beam has been established and extended to model a sailplane wing. The detailed procedure can be found in [8, 9]. The governing partial differential equations of motion for the beam with primes and over dots denoting partial differentiation with respect to position *y* and time *t* respectively are:

$$EIh'''' + m\ddot{h} - mx_\alpha\ddot{\psi} = 0 \quad (1)$$

$$GJ\psi'' + mx_\alpha\ddot{h} - I_\alpha\ddot{\psi} = 0 \quad (2)$$

For harmonic oscillation with circular frequency ω, the above differential equations can be solved using standard procedure [8, 9] to provide the solution for the amplitudes of both the bending displacement (H) and the torsional rotation (Ψ) in terms of six integration constants A_1 *to* A_6 as:

$$H(\xi) = A_1 cosh\alpha\xi + A_2 sinh\alpha\xi + A_3 cos\beta\xi + A_4 sin\beta\xi + A_5 cos\gamma\xi + A_6 sin\gamma\xi \quad (3)$$

$$\Psi(\xi) = A_1 k_\alpha cosh\alpha\xi + A_2 k_\alpha sinh\alpha\xi + A_3 k_\beta cos\beta\xi + A_4 k_\beta sin\beta\xi + A_5 k_\gamma cos\gamma\xi + A_6 k_\gamma sin\gamma\xi \quad (4)$$

where ξ is the non-dimensional length defined as $\xi = \frac{y}{L}$, and

$$\alpha = \left[2\left(\frac{q}{3}\right)^{\frac{1}{2}} \cos\left(\frac{\phi}{3}\right) - \frac{a}{3}\right]^{\frac{1}{2}},\ \beta = \left[2\left(\frac{q}{3}\right)^{\frac{1}{2}} \cos\left(\frac{(\pi-\phi)}{3}\right) + \frac{a}{3}\right]^{\frac{1}{2}},\ \gamma = \left[2\left(\frac{q}{3}\right)^{\frac{1}{2}} \cos\left(\frac{(\pi+\phi)}{3}\right) + \frac{a}{3}\right]^{\frac{1}{2}} \quad (5)$$

$$k_\alpha = \frac{b-\alpha^4}{bx_\alpha},\quad k_\beta = \frac{b-\beta^4}{bx_\alpha},\quad k_\gamma = \frac{b-\gamma^4}{bx_\alpha} \quad (6)$$

$$q = b + \frac{a^2}{3},\ \phi = cos^{-1}\left[\frac{27abc-9ab-2a^3}{\left\{2(a^2+3b)^{\frac{3}{2}}\right\}}\right],\ a = \left(\frac{I_\alpha\omega^2L^2}{GJ}\right),\ b = \left(\frac{m\omega^2L^4}{EI}\right),\ c = \left(\frac{I_\alpha - mx_\alpha^2}{I_\alpha}\right) \quad (7)$$

The expressions for bending rotation $\theta(\xi)$, bending moment $M(\xi)$, shear force $S(\xi)$ and torque $T(\xi)$ are given by [8, 9]

$$\theta(\xi) = \left(\frac{1}{L}\right)\{A_1\alpha sinh\alpha\xi + A_2\alpha cosh\alpha\xi - A_3\beta sin\beta\xi + A_4\beta cos\beta\xi - A_5\gamma sin\gamma\xi + A_6\gamma cos\gamma\xi\} \quad (8)$$

$$M(\xi) = -\left(\frac{EI}{L^2}\right)\{A_1\alpha^2 cosh\alpha\xi + A_2\alpha^2 sinh\alpha\xi - A_3\beta^2 cos\beta\xi - A_4\beta^2 sin\beta\xi - A_5\gamma^2 cos\gamma\xi - A_6\gamma^2 sin\gamma\xi\} \quad (9)$$

$$S(\xi) = \left(\frac{EI}{L^3}\right)\{A_1\alpha^3 sinh\alpha\xi + A_2\alpha^3 cosh\alpha\xi + A_3\beta^3 sin\beta\xi - A_4\beta^3 cos\beta\xi + A_5\gamma^3 sin\gamma\xi - A_6\gamma^3 cos\gamma\xi\} \quad (10)$$

$$T(\xi) = \left(\frac{GJ}{L}\right)\{A_1 k_\alpha\alpha sinh\alpha\xi + A_2 k_\alpha\alpha cosh\alpha\xi - A_3 k_\beta\beta sin\beta\xi + A_4 k_\beta\beta cos\beta\xi - A_5 k_\gamma\gamma sin\gamma\xi + A_6 k_\gamma\gamma cos\gamma\xi\} \quad (11)$$

Referring to figure 2, the boundary conditions for displacements are:

At $y = 0$ ($\xi = 0$): $H = H_1$, $\theta = \theta_1$, $\Psi = \Psi_1$ (12)

At $y = L$ ($\xi = 1$): $H = H_2$, $\theta = \theta_2$, $\Psi = \Psi_2$ (13)

IOP Conf. Series: Journal of Physics: Conf. Series **1048** (2018) 012010 doi:10.1088/1742-6596/1048/1/012010

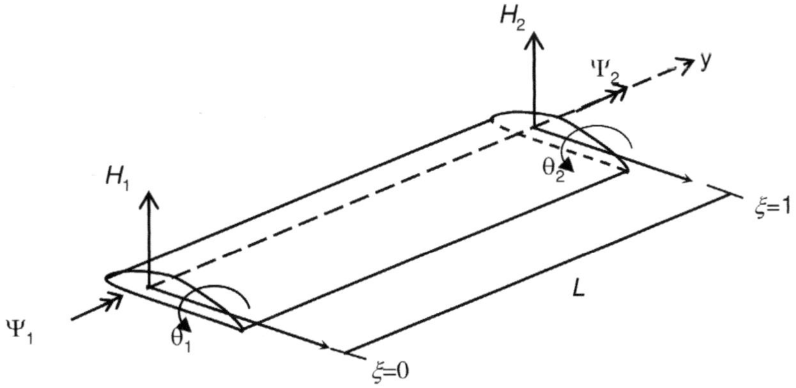

Figure 2. Boundary conditions for displacements of a sailplane wing element

Similarly referring to figure 3, the boundary conditions for the forces are:

At $y = 0$ (ξ=0): $S = S_1$, $M = M_1$, $T = -T_1$ (14)

At $y = L$ (ξ = 1): $H = -S_2$, $M = -M_2$, $T = T_2$ (15)

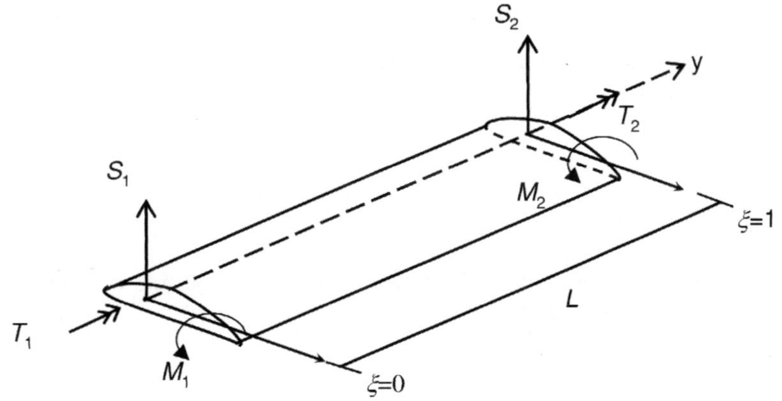

Figure 3. Boundary conditions for forces of a sailplane wing element

The expressions for displacements at the ends of the beam element can be obtained by applying the boundary conditions of equations (12) and (13) to equations (3), (4) and (8) to provide the following matrix relationship as:

$$\begin{bmatrix} H_1 \\ \theta_1 \\ \Psi_1 \\ H_2 \\ \theta_2 \\ \Psi_2 \end{bmatrix} = \begin{bmatrix} 1 & 0 & 1 & 0 & 1 & 0 \\ 0 & \alpha / L & 0 & \beta / L & 0 & \gamma / L \\ k_\alpha & 0 & k_\beta & 0 & k_\gamma & 0 \\ C_{h_\alpha} & S_{h_\alpha} & C_\beta & S_\beta & C_\gamma & S_\gamma \\ \alpha S_{h_\alpha} / L & \alpha C_{h_\alpha} / L & -\beta S_\beta / L & \beta C_\beta / L & -\gamma S_\gamma / L & \gamma C_\gamma / L \\ k_\alpha C_{h_\alpha} & k_\alpha S_{h_\alpha} & k_\beta C_\beta & k_\beta S_\beta & k_\gamma C_\gamma & k_\gamma S_\gamma \end{bmatrix} \begin{bmatrix} A_1 \\ A_2 \\ A_3 \\ A_4 \\ A_5 \\ A_6 \end{bmatrix} \quad (16)$$

or in matrix format as:

$$\mathbf{\Delta} = \mathbf{BA} \tag{17}$$

where **A** is the contact vector comprising the constants A_1 - A_6 and

$$C_{h_\alpha} = \cosh\alpha;\ S_{h_\alpha} = \sinh\alpha;\ C_\beta = \cos\beta;\ S_\beta = \sin\beta;\ C_\gamma = \cos\gamma;\ S_\gamma = \sin\gamma \tag{18}$$

The following matrix relationship can be obtained by substituting the boundary conditions of equations (14) and (15) for shear force, bending moment and torque into equations (9) - (11):

$$\begin{bmatrix} S_1 \\ M_1 \\ T_1 \\ S_2 \\ M_2 \\ T_2 \end{bmatrix} = \begin{bmatrix} 0 & W_3\alpha^3 & 0 & -W_3\beta^3 & 0 & -W_3\gamma^3 \\ -W_2\alpha^2 & 0 & W_2\beta^2 & 0 & W_2\gamma^2 & 0 \\ 0 & -W_1k_\alpha\alpha & 0 & -W_1k_\beta\beta & 0 & -W_1k_\gamma\gamma \\ -W_3\alpha^3S_{h_\alpha} & -W_3\alpha^3C_{h_\alpha} & -W_3\beta^3S_\beta & W_3\beta^3C_\beta & -W_3\gamma^3S_\gamma & W_3\gamma^3C_\gamma \\ W_2\alpha^2C_{h_\alpha} & W_2\alpha^2S_{h_\alpha} & -W_2\beta^2C_\beta & -W_2\beta^2S_\beta & -W_2\gamma^2C_\gamma & -W_2\gamma^2S_\gamma \\ W_1k_\alpha\alpha S_{h_\alpha} & W_1k_\alpha\alpha C_{h_\alpha} & -W_1k_\beta\beta S_\beta & W_1k_\beta\beta C_\beta & -W_1k_\gamma\gamma S_\gamma & W_1k_\gamma\gamma C_\gamma \end{bmatrix} \begin{bmatrix} A_1 \\ A_2 \\ A_3 \\ A_4 \\ A_5 \\ A_6 \end{bmatrix} \tag{19}$$

or in matrix format as:

$$\mathbf{F} = \mathbf{DA} \tag{20}$$

where

$$W_1 = \frac{GJ}{L},\ W_2 = \frac{EI}{L^2},\ W_3 = \frac{EI}{L^3} \tag{21}$$

The constant vector **A** can now be eliminated from equations (17) and (20) to give the following force-displacement relationship:

$$\mathbf{F} = \mathbf{K}\,\mathbf{\Delta} \tag{22}$$

where **K** is the 6×6 frequency dependent dynamic stiffness matrix given by

$$\mathbf{K} = \mathbf{D}\,\mathbf{B}^{-1} \tag{23}$$

The dynamic stiffness matrix in equation (23) representing a bending-torsion coupled beam such as an aircraft wing can now be used to model an aircraft wing.

2.2 *Application of the Wittrick-Williams algorithm*

The dynamic stiffness matrix in equation (23) can be used to compute the natural frequencies and mode shapes of a bending-torsion coupled beam. A non-uniform aircraft wing can be modelled as an assembly of many uniform bending-torsion coupled beams. For example, a non-uniform wing can be modelled as a stepped wing where it is split into a number of uniform elements. The dynamic stiffness matrices of all individual elements are assembled to form the overall dynamic stiffness matrix $\mathbf{K}_f$ of the complete wing. The natural frequency computation is accomplished by applying the Wittrick-Williams algorithm [7]. The algorithm monitors the Sturm sequence condition of $\mathbf{K}_f$ in such a way that there is no possibility of missing any natural frequency of the wing. The application procedure of the algorithm is briefly summarised here. Supposing that ω denotes the angular frequency of the vibrating wing, the number of natural frequencies passed (j) as ω is increased from zero to ω^* is given by

$$j = j_0 + s\{\mathbf{K}_f\} \tag{24}$$

where $\mathbf{K}_f$, the overall dynamic stiffness matrix of the wing whose elements depend on ω is evaluated at $\omega = \omega^*$; $s\{\mathbf{K}_f\}$ is the number of negative elements on the leading diagonal of $\mathbf{K}_f^\Delta$, $\mathbf{K}_f^\Delta$ is the upper triangular matrix obtained by applying the usual form of Gauss elimination to $\mathbf{K}_f$, and j_0 is the number of natural frequencies of the wing still lying between 0 and ω^* when the displacement components to which $\mathbf{K}_f$ corresponds are all zeros.(Note that the structure can still have natural frequencies when all its nodes are clamped, because exact member equations allow each individual member to displace between nodes with an infinite number of degrees of freedom, and hence infinite number of natural frequencies between nodes. Thus

$$j_0 = \sum j_m \tag{25}$$

where j_m is the number of natural frequencies between 0 and ω^* for an individual component member with its ends fully clamped, while the summation extends over all members of the structure. Thus, with equations (24) and (25), it is possible to ascertain how many natural frequencies of the wing lie below an arbitrarily chosen trial frequency (ω^*). This feature of the algorithm can be used to converge upon any required natural frequency to any desired accuracy. As successive trial frequencies can be chosen, computer implementation of the algorithm is simple. For a detailed explanation and understanding of the Wittrick-Williams algorithm, readers are referred to the original work of Wittrick and Williams [7].

3. Results and discussion

Two sailplane wings with cantilever boundary conditions are analysed for their modal characteristics. The sailplanes are named as S_1 and S_2 to preserve commercial anonymity. The particulars for the two sailplanes are given in table 1. The unit for each parameter is provided in parentheses in the table.

Table 1. Particulars of the two sailplanes.

Parameters	Sailplane-S_1	Sailplane-S_2
Wing Span (m)	22	15
Wing Area (m^2)	15.44	10.05
Aspect Ratio	31.35	22.4
Wing Root Chord (m)	1.0	0.9
Wing Tip Chord (m)	0.4	0.4
Sweep angle (deg)	0	0
Fuselage Length (m)	7.6	6.72
Height Overall (m)	2.0	2.0
Weight Empty (kg)	390	234
Max Take-off weight (kg)	550	440
Max Wing Loading (kg/m^2)	37	36
Max Cruising Speed (knots)	135	105

A computer program in Fortran was developed to compute the natural frequencies and mode shapes of the wings using the dynamic stiffness method and the Wittrick-Williams algorithm. The first five natural frequencies and mode shapes of the two sailplanes are presented in table 2 and figure 4, respectively. In figure 4, the horizontal axis is the non-dimensional length (ξ) and the vertical axis is bending displacement (H) and/or the torsional rotation (Ψ). The letters B and T in table 2 indicate bending and torsion dominated modes, respectively whereas the letter C indicates a coupled mode which has substantial amount of both bending and torsional displacements.

Table 2. The natural frequencies of the two sailplane wings.

Category	Natural Frequencies (ω_i) (rad/s)				
	ω_1	ω_2	ω_3	ω_4	ω_5
Sailplane S1	10.64 (B)	42.62 (B)	109.6 (B)	111.5 (T)	201.4 (B)
Sailplane S2	13.38 (B)	42.09 (B)	93.35 (B)	164.2 (T)	167.4 (C)

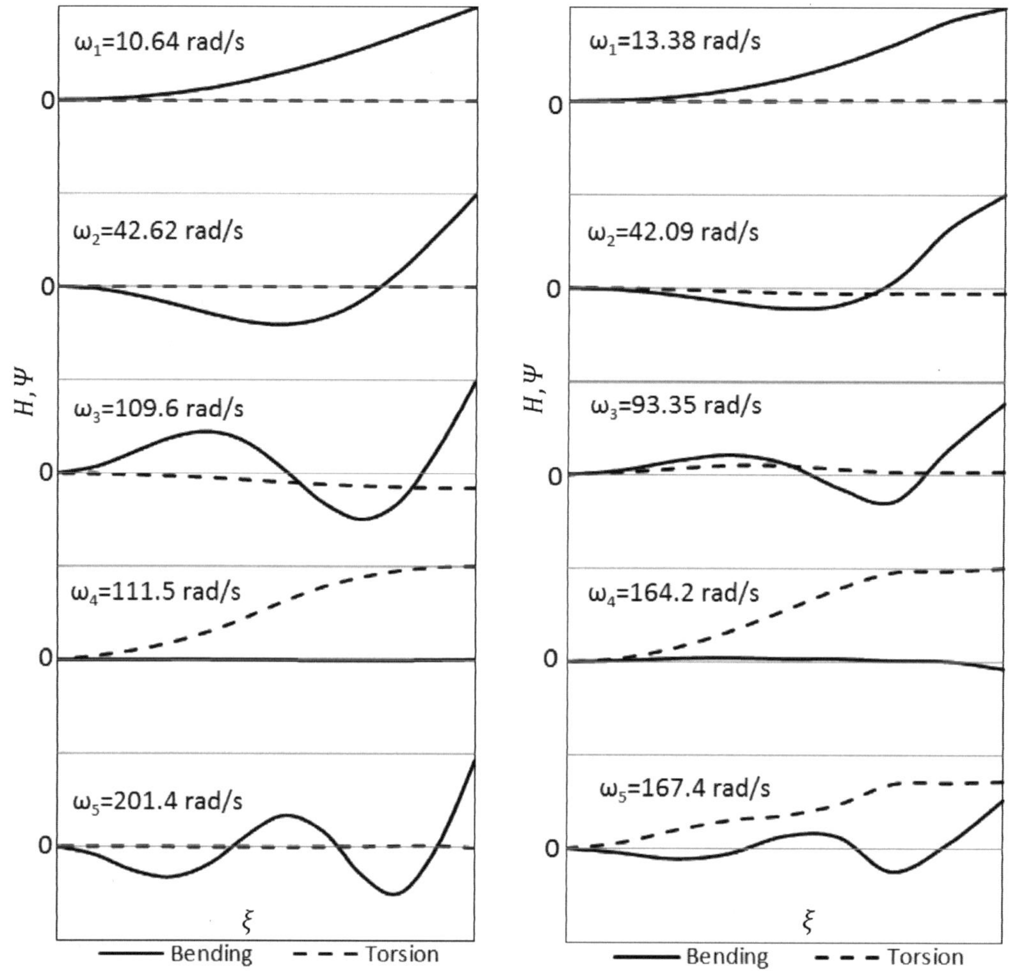

(a) Sailplane S1 (b) Sailplane S2

Figure 4. Natural frequencies and mode shapes of Sailplane wings S1 and S2.

A detailed parametric study with the variation of bending and torsional rigidities is then carried out. The bending (*EI*) and torsional (*GJ*) rigidities for each wing are varied between -25% to 25% with an increment of 5% in each step. Their subsequent effects on the natural frequencies and mode shapes are investigated. The first six natural frequencies with the variation of *EI* and *GJ* for S_1 are presented in tables 3 and 4 respectively. Clearly in table 3, all the natural frequencies corresponding to bending modes increase with the variation of the bending rigidity *EI* from -25% to +25% whereas the 6th one remains unchanged because it is a high frequency torsional mode which is not expected to change greatly as the torsional rigidity *GJ* is kept constant. Interesting results can be found in that there are modal interchanges (flip-over) for the variation of *EI* as the mode shapes changes from bending to torsional mode, see the 3rd natural frequency in table 3. Similar mode shape changes from torsional to bending mode can be found, for example in the 4th mode of table 3. With respect to the variation of *GJ* shown in table 4, the 1st, 2nd and 5th natural frequencies remain unchanged whereas the rest increase. Similar pattern can be observed that the 3rd and 4th modes are changed from torsional to bending mode and bending to torsion mode, respectively.

Table 3. The effects on the natural frequencies for S_1 with the variation of *EI*.

Variation in *EI* (%)	ω_i (rad/s)					
	ω_1 (B)	ω_2 (B)	ω_3	ω_4	ω_5 (B)	ω_6 (T)
-25	9.217	36.91	94.96 (B)	111.5 (T)	174.4	260.8
-20	9.519	38.12	98.07 (B)	111.5 (T)	180.1	260.8
-15	9.811	39.29	101.1 (B)	111.5 (T)	185.7	260.8
-10	10.10	40.43	104.0 (B)	111.5 (T)	191.1	260.8
-5	10.37	41.54	106.9 (B)	111.5 (T)	196.3	260.8
0	10.64	42.62	109.6 (B)	111.5 (T)	201.4	260.8
5	10.91	43.67	111.5 (T)	112.4 (B)	206.4	260.8
10	11.16	44.70	111.5 (T)	115.0 (B)	211.2	260.8
15	11.41	45.71	111.5 (T)	117.6 (B)	216.0	260.8
20	11.66	46.69	111.5 (T)	120.1 (B)	220.6	260.8
25	11.90	47.65	111.5 (T)	122.6 (B)	225.2	260.8

Table 4. The effects on the natural frequencies for S_1 with the variation of *GJ*.

Variation in *GJ* (%)	ω_i (rad/s)					
	ω_1 (B)	ω_2 (B)	ω_3	ω_4	ω_5 (B)	ω_6 (T)
-25	10.64	42.62	96.59 (T)	109.6 (B)	201.4	225.9
-20	10.64	42.62	99.76 (T)	109.6 (B)	201.4	233.3
-15	10.64	42.62	102.8 (T)	109.6 (B)	201.4	240.5
-10	10.64	42.62	105.8 (T)	109.6 (B)	201.4	247.4
-5	10.64	42.62	108.7 (T)	109.6 (B)	201.4	254.2
0	10.64	42.62	109.6 (B)	111.5 (T)	201.4	260.8
5	10.64	42.62	109.6 (B)	114.3 (T)	201.4	267.2
10	10.64	42.62	109.6 (B)	117.0 (T)	201.4	273.5
15	10.64	42.62	109.6 (B)	119.6 (T)	201.4	279.7
20	10.64	42.62	109.6 (B)	122.2 (T)	201.4	285.7
25	10.64	42.62	109.6 (B)	124.7 (T)	201.4	291.6

For the sailplane S_2, the first six natural frequencies with the variation of *EI* and *GJ* are presented in tables 5 and 6 respectively. Similar patterns, like the sailplane S_1, can be noted in respect of the modal interchanges (flip-overs), see the 4th and 5th modes.

It is well known that in flutter and response analysis of aeronautical and other structures, natural frequencies and mode shapes arising from the free vibration analysis play a very important role, particularly when the normal mode method is used. Essentially the mode shapes are appropriately scaled by the corresponding generalised coordinates when computing the flutter mode or the overall response of a structure such as an aircraft wing. The mode shapes provide essential information concerning the dynamic behaviour of a structural system and they are useful indicators of the properties of the system. It is also important to note that some of the modes can couple with each other which can give rise to instability, resulting in structural resonance causing eventual failure of the structure. In aeronautical applications, the presence of the aerodynamic forces alters the mode shapes significantly, triggering frequency coalescence phenomenon for which the classical bending-torsion flutter is a typical example. The importance of natural frequency and mode shape calculation cannot be overemphasised.

Table 5. The first six natural frequencies for S_2 with the variation of EI.

Variation in EI (%)	ω_i (rad/s)					
	ω_1 (B)	ω_2 (B)	ω_3 (B)	ω_4	ω_5	ω_6 (C)
-25	11.64	36.51	81.19	145.8 (B)	164.3 (T)	251.8
-20	11.97	37.69	83.66	149.9 (B)	164.3 (T)	259.3
-15	12.34	38.84	86.20	154.4 (B)	164.4 (T)	267.0
-10	12.69	39.96	88.65	158.7 (B)	164.5 (T)	274.5
-5	13.04	41.04	91.04	162.6 (B)	164.9 (T)	281.6
0	13.38	42.10	93.35	164.2 (T)	167.4 (C)	288.6
5	13.71	43.12	95.60	164.4 (T)	171.2 (C)	295.3
10	14.03	44.13	97.81	164.5 (T)	175.1 (C)	301.9
15	14.34	45.11	99.95	164.6 (T)	178.8 (C)	308.3
20	14.65	46.07	102.0	164.7 (T)	182.5 (C)	314.6
25	14.95	47.01	104.1	164.7 (T)	186.1 (C)	320.6

Table 6. The first six natural frequencies for S_2 with the variation of GJ.

Variation in GJ (%)	ω_i (rad/s)					
	ω_1 (B)	ω_2 (B)	ω_3 (B)	ω_4	ω_5	ω_6 (C)
-25	13.37	42.03	93.01	142.8 (T)	166.2 (C)	286.2
-20	13.38	42.04	93.10	147.3 (T)	166.5 (C)	286.8
-15	13.38	42.06	93.18	151.8 (T)	166.7 (C)	287.3
-10	13.38	42.07	93.24	156.1 (T)	166.9 (C)	287.8
-5	13.38	42.09	93.30	160.3 (T)	167.1 (C)	288.2
0	13.38	42.10	93.35	164.2 (T)	167.4 (C)	288.6
5	13.38	42.11	93.41	166.9 (C)	171.2 (T)	289.0
10	13.38	42.11	93.44	167.2 (C)	172.6 (T)	289.2
15	13.38	42.12	93.48	167.4 (C)	176.4 (T)	289.5
20	13.38	42.13	93.51	167.5 (C)	180.1 (T)	289.7
25	13.38	42.14	93.54	167.6 (C)	183.7 (T)	289.9

4. Conclusions

Using the dynamic stiffness method together with the Wittrick-Williams algorithm as the solution technique, the modal behaviour of two sailplane wings is investigated. Natural frequencies and mode shapes for these wings are presented and the results are examined and discussed. The bending and torsional rigidities of each wing are then varied and their subsequent effects on the natural frequencies and mode shapes are investigated. A detailed parametric study with the variations of bending and torsional rigidities provides some interesting results showing modal interchanges which can be of practical help in the design of such wings. The investigation paves the way for further research to establish trends for the modal behaviour of high aspect ratio sailplane wings.

References

[1] Banerjee JR 1984 *J. Aircraft* **21(9)** 733-5.
[2] Banerjee JR 1988 *J. Aircraft* **25(5)** 473-6.
[3] van Schoor M C and van Flotow AH 1990 *J. Aircraft* **27(10)** 901-8.
[4] Eslimy-Isfahany SHR, Banerjee JR and Sobey AJ 1996 *J. Sound and Vibration* **195(2)** 267-83.
[5] Tang D and Dowell EH 2001 AIAA **39(8)** 1430-41.
[6] Banerjee JR, Liu X and Kassem HI 2014 *J. Applied Nonlinear Dynamics*, **3(4)** 413-22.

IOP Conf. Series: Journal of Physics: Conf. Series **1048** (2018) 012010 doi:10.1088/1742-6596/1048/1/012010

[7] Wittrick WH and Williams FW 1971 Quarterly J. Mech;.and App. Mathematics **24(3)** 263-84.
[8] Banerjee JR 1989 Int. J. Numerical Methods in Engineering **28 (6)** 1283-98.
[9] Banerjee JR 1991 Adv. in Engineering Software, **13(1)** 17-24.

IOP Conf. Series: Journal of Physics: Conf. Series **1048** (2018) 012011 doi:10.1088/1742-6596/1048/1/012011

Energy efficient active vibration control strategies using electromagnetic linear actuators

Angel Torres-Perez, Ali Hassan, Stefan Kaczmarczyk, Phil Picton

The University of Northampton, St. George's Avenue, Northampton, NN2 6JD, United Kingdom

E-mail: angel.torresperez@northampton.ac.uk

Abstract. Energy efficient current control methods in electromagnetic linear actuators are required to minimize the electrical power requirements imposed by active vibration control strategies. In this paper an efficient bidirectional buck-boost converter is discussed in two scenarios: an active vibration isolation system and an active dynamic vibration absorber (ADVA) using a voice coil motor (VCM) actuator. An electrical analogous circuit of an experimental test platform is used as part of the simulation model. This test platform is based on a vibration shaker that provides the based excitation required for the single Degree of-Freedom (1DoF) vibration model under study. The proposed bidirectional non-isolated buck-boost converter can recover the energy when the VCM acts as a generator and store it for future use. Simulation results prove that this type of topology is far more efficient than linear amplifiers typically used in active vibration control. Within the context of slender structures, this efficient current control method improves the viability of using active vibration control in flexible structures such as beams.

Keywords: active DVA, bidirectional DC converters, switching power amplifier, buck-boost, voice coil motors

1. Introduction

A dynamic vibration absorber (DVA), or vibration neutralizer is a tuned spring-mass system which reduces or eliminates the vibration of a harmonically excited system. There are two types of DVAs depending on the need of external power to perform their functions. Passive DVAs do not require actuators and they cannot guarantee high vibration control performance for disturbance with various frequencies in general. However, active dynamic vibration absorbers (ADVAs) have sensors and actuators and are often considered when high vibration control performance is required for disturbance with various frequencies. The actuators of ADVAs are typically controlled using feedforward or close loop controllers. DVAs has been extensively studied in building structures, as defence mechanism against earthquakes, to counter seismic movements and wind forces [1].

In this paper an electrodynamic actuator is used to modify and effectively control the mechanical stiffness and damping of a tuned mass damper. A feedforward controller is used to ensure the natural frequency of the vibration absorber tracks the disturbance main excitation frequency. As a result, the system has optimal performance throughout a large range of excitation frequencies. The feedforward controller requires accurate current control flowing through the electrodynamic actuator.

This study analyses an active DVA using an electrical analogous circuit model as discussed in [2]. This electrical analogy is used to set a comparison between a linear bridge current control amplifier and a switch mode power amplifier (class D) based on a bidirectional buck-boost converter.

2. Methodology

2.1. Active control rig

The experimental rig shown in Figure 1a. was built to test the concept of active control using an adjustable mass connected to a planar spring and one VCM actuator. Two motion sensors (piezoelectric accelerometers) are connected to the absorber mass and the shaker table respectively. Two electrical signals are recorded for VCM characterization that correspond to the current and voltage signals which are measured using one isolated differential voltage probe and one current probe respectively. The physical model which is a pictorial representation of the system under study is shown in Figure 1b.

The vibration model corresponds to a 1DoF with base excitation as shown in Figure 2.

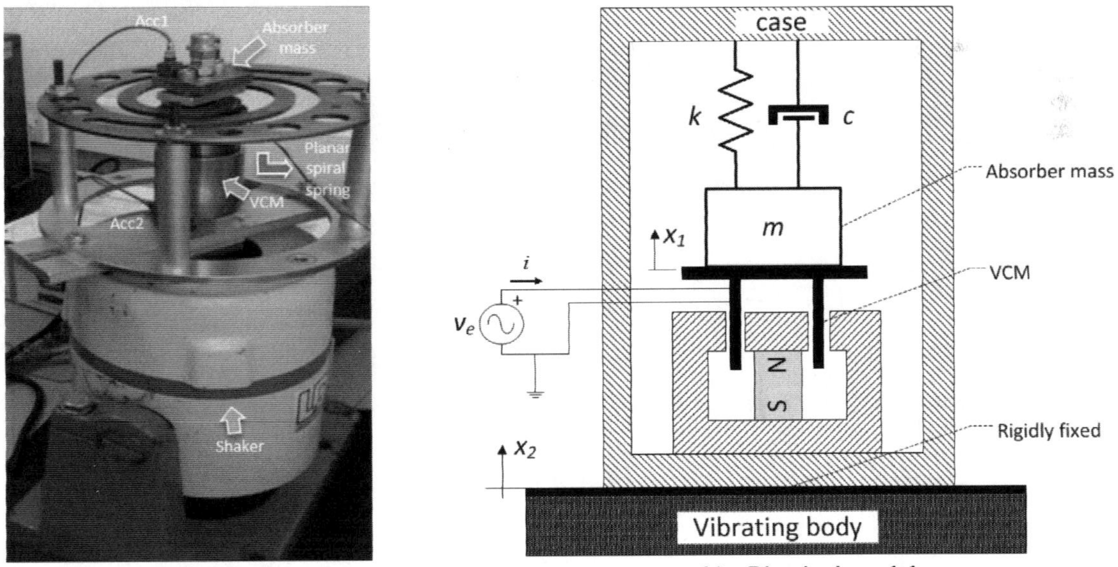

a) Experimental rig b) Physical model

Figure 1. Active vibration control rig [2]

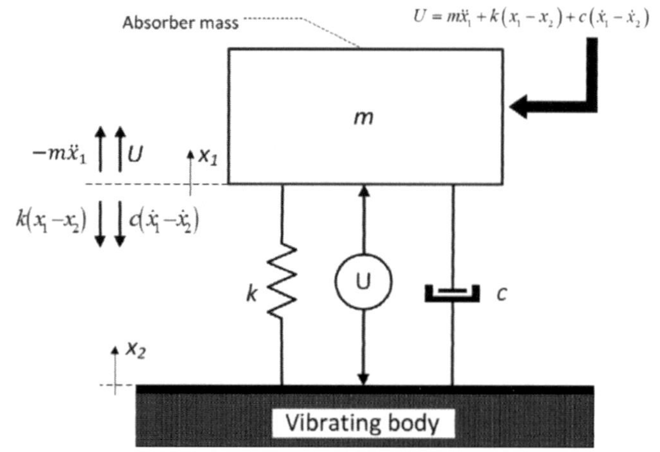

Figure 2. Active control vibration model with a generic actuator U (1DoF) [2]

2.2. *Electromechanical Analogous Circuit*

If the type of actuator is a Voice Coil Motor, the coupling equation between the electrical and mechanical system is due to Lorentz Force and Faradays's Law of Induction. The actuator mechanical force is proportional to actuator's current when there is a constant magnetic flux density in the VCM gap under all actuator stroke positions $N\frac{d\Phi}{dx} = cte$ as shown in (1).

$$F = \beta l \cdot i = \alpha_{VCM} \cdot i \tag{1}$$

The constant that multiplies the actuator current is referred as motor constant α_{VCM} . The definition is shown in (2) where α_{VCM} is the motor constant; N is the number of turns; β the magnetic flux density in the gap and l the length of the wounded wire.

$$\alpha_{VCM} = N\frac{d\Phi}{dx} = \beta l \tag{2}$$

The Back Electromotive Force is proportional to the total number of turns and actuator speed as shown in (3).

$$E = -N\frac{d\Phi}{dt} = -N\frac{d\Phi}{dx}\frac{dx}{dt} = -\alpha_{VCM}\dot{x} \tag{3}$$

A possible circuit representation that couples the mechanical system and electrical side uses a gyrator and it is shown in Figure 3. The governing differential equations for this circuit are obtained by applying Kirchoff's Voltage Law in loop1 and loop 2 as shown in (4).

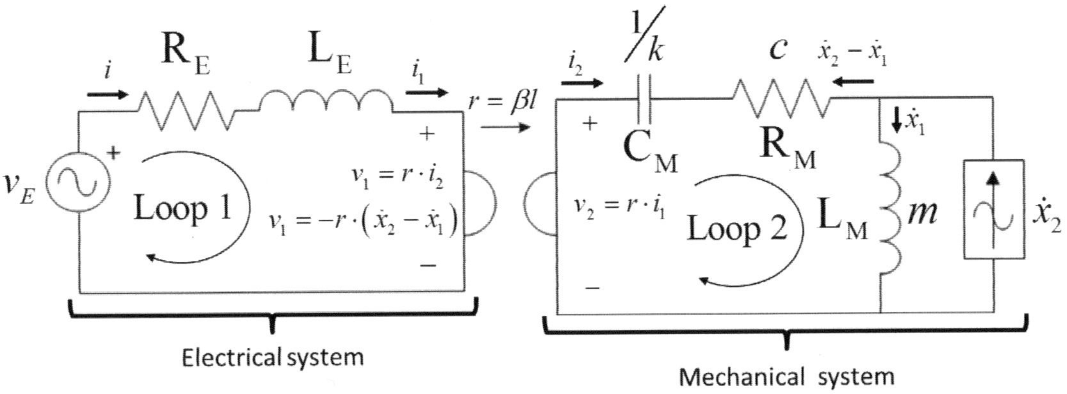

Figure 3. 1DoF active control rig with VCM (Type II)

$$\begin{aligned} &V_E - R_E \cdot i - L_E \cdot \frac{di}{dt} - v_1 = 0 \quad \Big| \; i = i_1 \\ &v_2 + k(x_2 - x_1) + c(\dot{x}_2 - \dot{x}_1) - m\ddot{x}_1 = 0 \quad \Big| \; i_2 = -(\dot{x}_2 - \dot{x}_1) \end{aligned} \tag{4}$$

These equations could be represented as a function of the currents in the electrical and mechanical sides of Figure 3 as shown in (5). The relationship between electrical and mechanical quantities is shown in Table 1.

$$r = \beta l = \alpha_{VCM} \quad \begin{cases} V_E - R_E \cdot i - L_E \cdot \dfrac{di}{dt} + r \cdot (\dot{x}_2 - \dot{x}_1) = 0 \\ r \cdot i + k(x_2 - x_1) + c(\dot{x}_2 - \dot{x}_1) - m\ddot{x}_1 = 0 \quad \Big| \; U = r \cdot i = \alpha_{VCM} \cdot i \end{cases} \tag{5}$$

Table 1. Interpretation of electrical and mechanical quantities

Electrical Quantity	Mechanical Analog II (Force Voltage)
Voltage, e	Force, f
Current, i	Velocity, v
Resistance, R	Friction, B
Capacitance, C	Compliance, 1/K (Inverse spring constant)
Inductance, L	Mass, M

A better circuit representation that considers damping due to eddy currents in the VCM when there is an open circuit across the terminals is shown in Figure 4. The circuit shown in Figure 4 provides a trend that is in good agreement with experimental results when the VCM terminals are under open and closed circuit conditions respectively. The circuit values are shown in Table 2. A further discussion about the derivation of this circuit is shown in [2] . It has been observed that the damping due to eddy currents it is also affected by the actuator stroke and current flowing through the main coil.

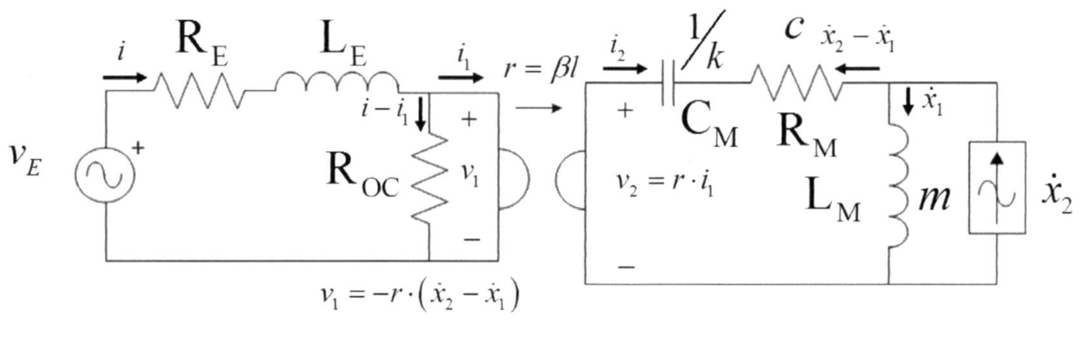

Figure 4. 1DoF active control rig with VCM

Table 2. Circuit values

Parameter	m	c	k	$f_0 = \frac{1}{2\pi}\sqrt{\frac{k}{m}}$	$\alpha_{VCM} = r$	
Value	432.4	0.5 Negligible	42.167	49.7	7.4	
Unit	[g]	[Ns/m]	[kN/m]	[Hz]	[N/A]	
Parameter	L_E	R_E	R_{OC}	C_M	L_M	R_M
Value	4.1	3.3	9.78	23.71	0.4324	0.5
Unit	[mH]	[Ω]	[Ω]	[µF]	[mH]	[Ω]

3. ACTIVE CONTROL LAWS (ADVA)

The force U introduced by the actuator using the feedforward controller as shown in (6) must remove the damping of the system *c* and modify the natural frequency of the absorber to be the same as the main harmonic component of the vibration signal [3].

$$U(j\omega) = V(j\omega)\left(c + j\omega\underbrace{\left(m - \frac{k}{\omega^2}\right)}_{\alpha}\right) = V(j\omega)\left(c + \frac{j}{\omega}\underbrace{\left(\omega^2 m - k\right)}_{\gamma}\right) \tag{6}$$
$$V(j\omega) = j\omega \cdot \left(X_1(j\omega) - X_2(j\omega)\right)$$

Where $V(j\omega)$ is the relative speed of the absorber mass with respect to the vibrating body in the frequency domain. The absorber tuning could be shifted to the harmonic excitation frequency ω by introducing a force proportional to actuator relative position $(x_1 - x_2)$ times a constant of magnitude α or a force proportional to actuator relative acceleration $(\ddot{x}_1 - \ddot{x}_2)$ times a constant of magnitude γ. Thus, two possible gains, one proportional to acceleration (α) and the other proportional to position (γ) could be defined relative to the natural resonant frequency of the absorber mass ω_0. These gains are shown in (7) and (8).

IOP Conf. Series: Journal of Physics: Conf. Series **1048** (2018) 012011 doi:10.1088/1742-6596/1048/1/012011

$$\alpha = \left(m - \frac{k}{\omega^2}\right) \xrightarrow{\omega_0 = \sqrt{\frac{k}{m}}} \alpha = m\left(1 - \left(\frac{\omega_0}{\omega}\right)^2\right) \quad (7)$$

$$\gamma = \left(\omega^2 m - k\right) \xrightarrow{\omega_0 = \sqrt{\frac{k}{m}}} \gamma = m\left(\omega^2 - \omega_0^2\right) \quad (8)$$

The force U, exerted by the VCM is the current times the motor constant (r). Therefore, the actuator force is controlled by the current flowing through the VCM terminals. A current controlled power amplifier is needed for this task. The stiffness in the mechanical side of Figure 4. could be modified by selecting a current *i* that is proportional to the relative position times the position gain over the motor constant γ/r. The effective mass in the mechanical side of Figure 4 could be altered by selecting a current that is proportional to the relative acceleration times the acceleration gain over the motor constant α/r.

4. CURRENT CONTROL AMPLIFIER

The current flowing through the VCM is adjusted using a power amplifier. The feedforward controller requires an amplifier with a fast-current loop minimizing any delays in the actuator force. This precision amplifier is typically implemented with a class AB amplifier with low distortion level, negligible output voltage/current ripple and poor efficiency (50%). The main drawback of this power amplifier is that the dissipated power depends on output power factor. This type of amplifier is quite inefficient when driving low impedance and reactive loads. All the reflected energy from reactive components and back EMF is consumed dissipating heat. A block diagram of a class AB amplifier with current control is shown in Figure 5a and in a bridge configuration in Figure 5b.

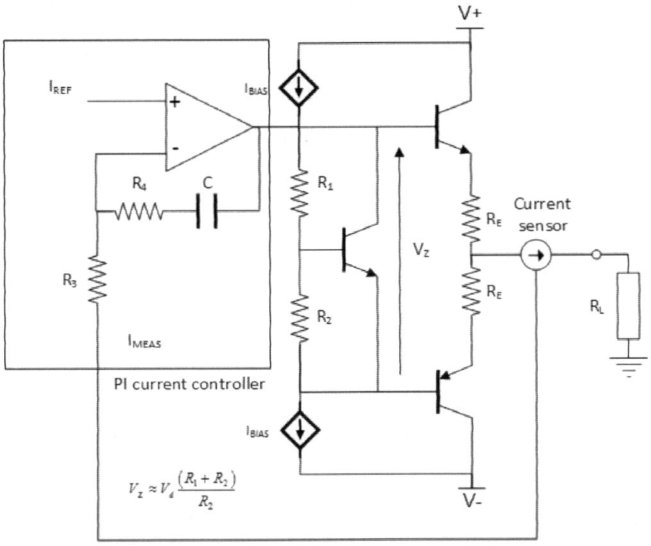

Figure 5a. Linear amplifier (class AB), push-pull output stage

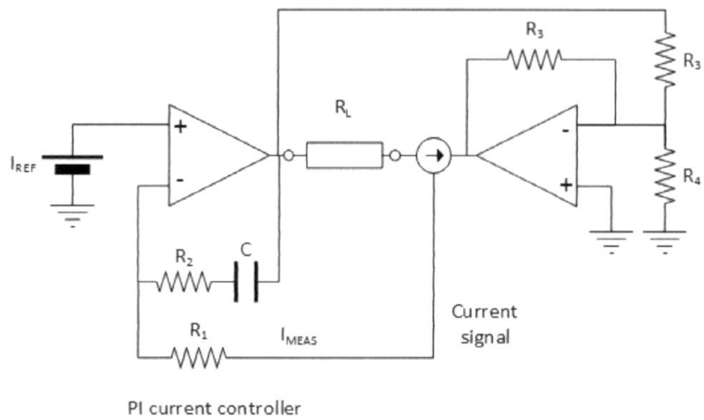

Figure 5b. Linear amplifier (class AB) in a bridge configuration

In this paper, an efficient current controlled class D amplifier controls the current through the VCM. The circuit topology in Figure 6a resembles a synchronous buck converter [4] and it is sometimes referred as bidirectional buck-boost converter [5]. A small signal model and average state space model is shown in [5]. The steady state analysis assumes a constant average current through the inductor, hence the volt-second balanced condition in one switching cycle is met and the ratio between the output and input voltages could be obtained as shown in (9) where Ts is the switching period and D is the duty cycle.

$$(V_s - V)DT_s = (V_s + V)(1-D)T_s \rightarrow \frac{V}{V_s} = D-(1-D) = 2D-1 \tag{9}$$

Thus, the DC voltage gain is a linear function of the duty cycle D as shown in Figure 6b. For D > 0.5 the output voltage is of positive polarity, while for D < 0.5 it is of negative polarity. The switch S could be implemented using two transistors allowing a bidirectional current flow.

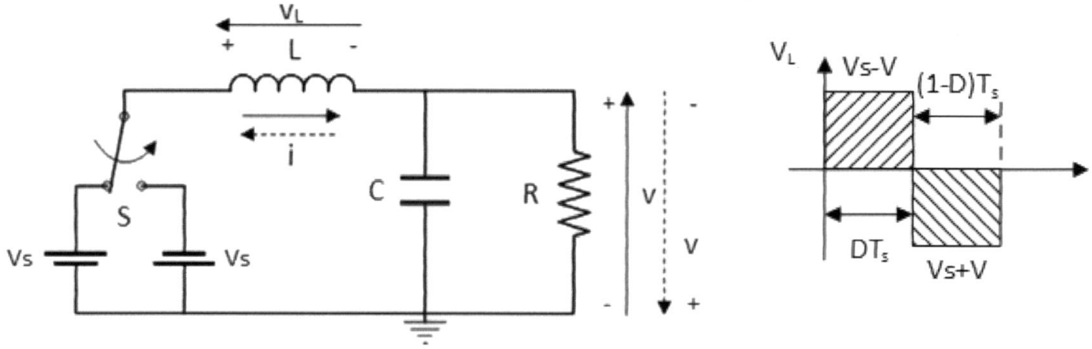

Figure 6a. Switching amplifier derived from the buck converter

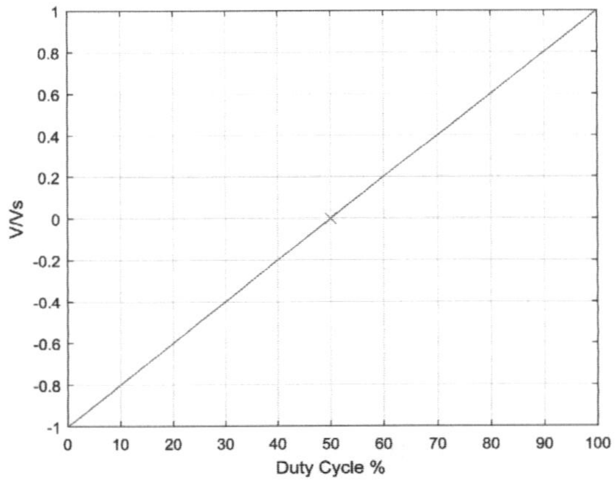

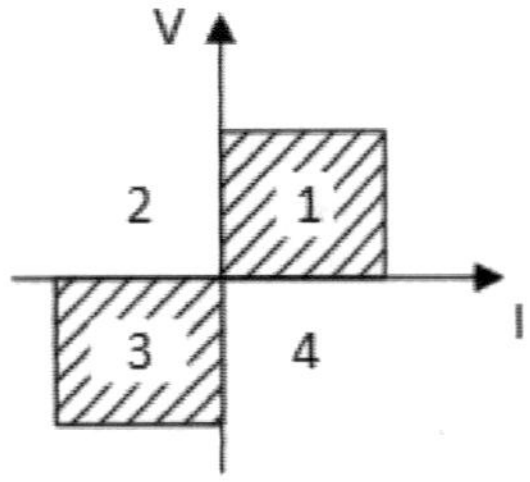

Figure 6b. Ratio between the output and input voltages (two quadrant operation)

The hardware implementation of the switch S in Figure 6a, corresponds to a power amplifier in a half bridge configuration as shown in Figure 7a. A single power supply implementation requires four transistors according to the bridge circuit in Figure 7b. The PWM output of the PI current controller in Figure 7c is the control signal for the transistors of the class D amplifier. The PWM signal is generated comparing the output signal of the PI controller against a high frequency sawtooth signal.

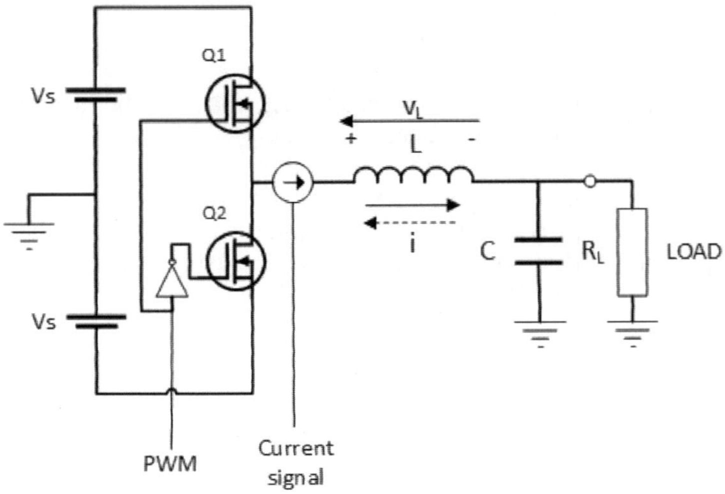

Figure 7a. Class D amplifier (half bridge)

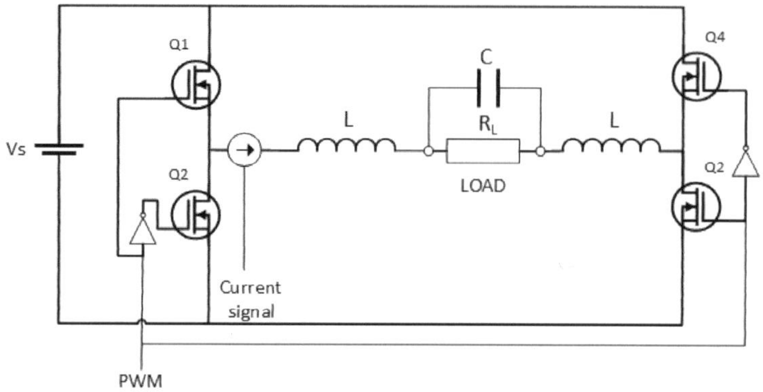

Figure 7b. Class D amplifier (full bridge)

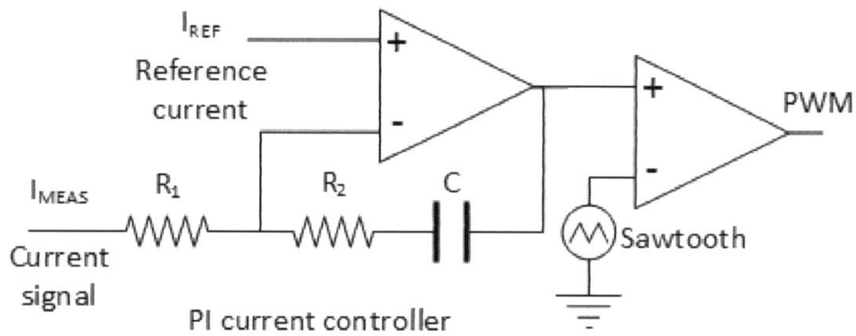

Figure 7c. PWM signal generation (current control)

5. Simulation results

The full bridge implementation of the switching amplifier with the analogous circuit as the load was simulated in PSIM. The low pass differential LC filter was designed with a cut-off frequency of 230Hz which is much lower than the PWM frequency. The inductor L is 1mH and the capacitor 470μF. The equivalent single ended filter shown in Figure 8 is used for calculating the cut off frequency and quality factor calculations of the differential filter. The base excitation (shaker table) was assumed to be a constant amplitude sinusoidal speed signal of 0.1m/s of peak amplitude and a frequency of 50Hz.

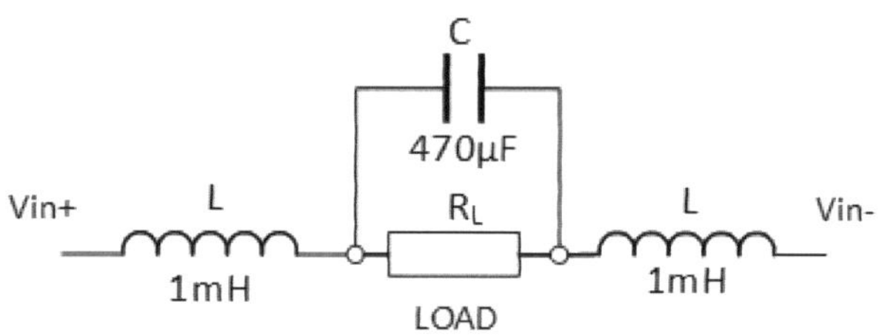

Figure 8. Differential filter and it is Single-Ended Equivalent

IOP Conf. Series: Journal of Physics: Conf. Series **1048** (2018) 012011 doi:10.1088/1742-6596/1048/1/012011

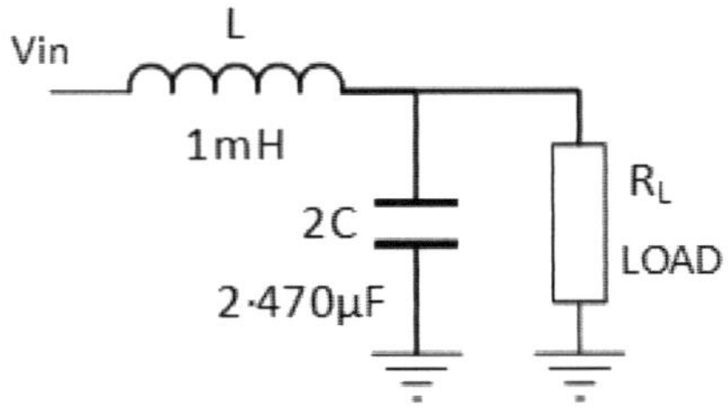

A frequency response analysis of the switching amplifier shown in Figures 9a-b was conducted in open loop around the operating point (D=50%). The gyrator block in Figure 9b was implemented with two current controlled voltage sources. The ac sweep was conducted by varying the duty cycle around the operating point (signal injection). The open loop response of the inductor current as a function of the duty cycle for different frequencies is shown in Figure 9c. The system is stable, and the phase margin is 90°.

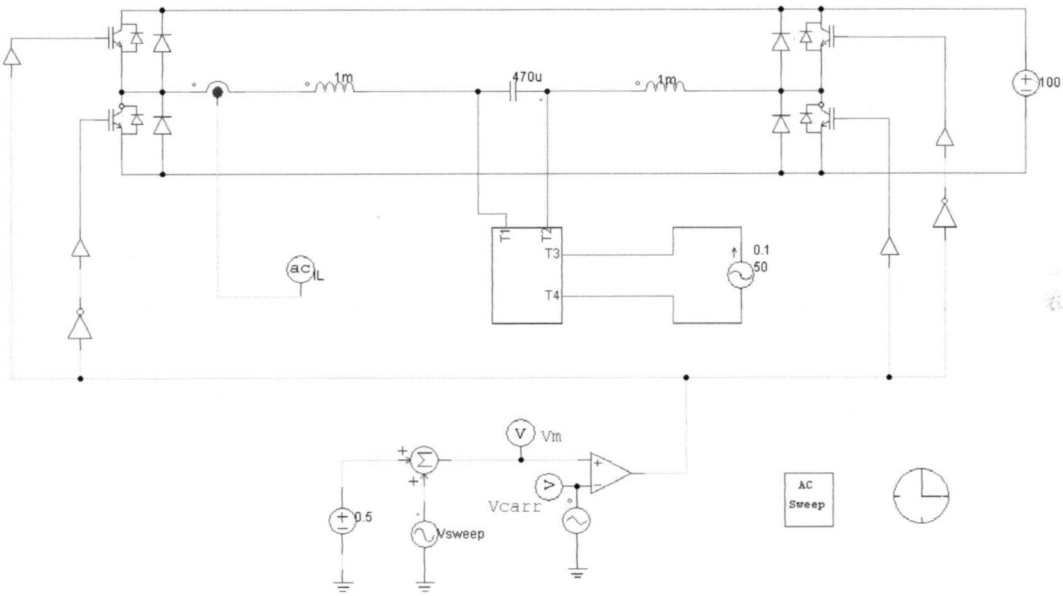

Figure 9a. Class D amplifier with analogous circuit as the load

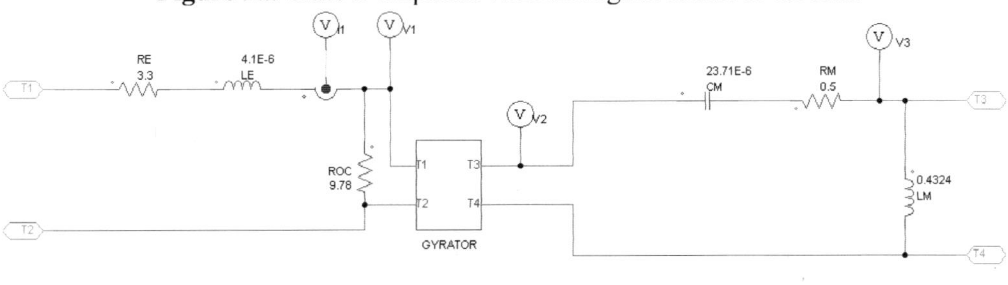

Figure 9b. Analogous circuit implementation (load)

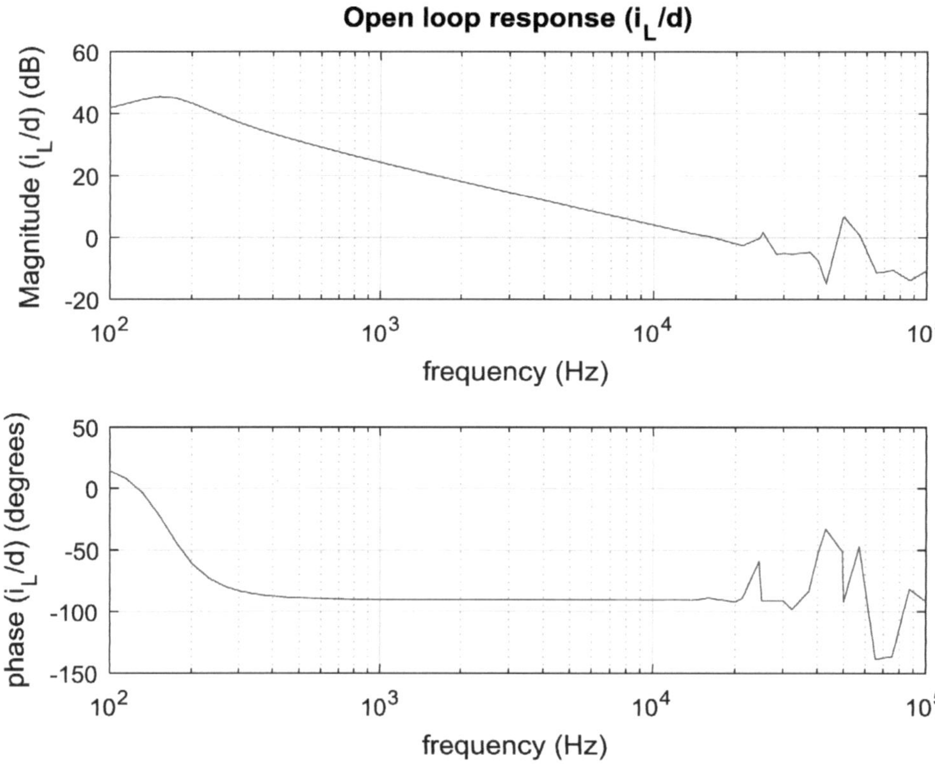

Figure 9c. AC Sweep analysis (open-loop, duty cycle 50%)

Closed loop response was analysed using the transient analysis shown in Figures 10a-c. Current control was performed using a PI regulator G(s) = k (1 + sT) / (sT) with a gain of 8 and a time constant of 100μs. The PWM signal is generated comparing the output of the PI controller against a sawtooth signal frequency of 50Khz and peak to peak voltage of 1V. The saturation block after the PI regulator limits the range of the duty cycle from 0 to to 95% (it corresponds with an output between 0 and 0.95V). In Figure 10b, a transient analysis was performed to measure power requirements and the differential voltage across the load. Figure 10c represents the measured and reference load currents when the reference current signal is a sine wave of 50Hz and 2A of peak voltage with 1A of DC offset. The simulation results in Figure 10b do not consider the on resistance of the switches, the resistance of the output filter and wire resistances. Therefore, the power conversion efficiency is ideal, and all the input power is transferred to the load. A linear amplifier (AB type) will never reach this efficiency because the transistors are operating in the linear region.

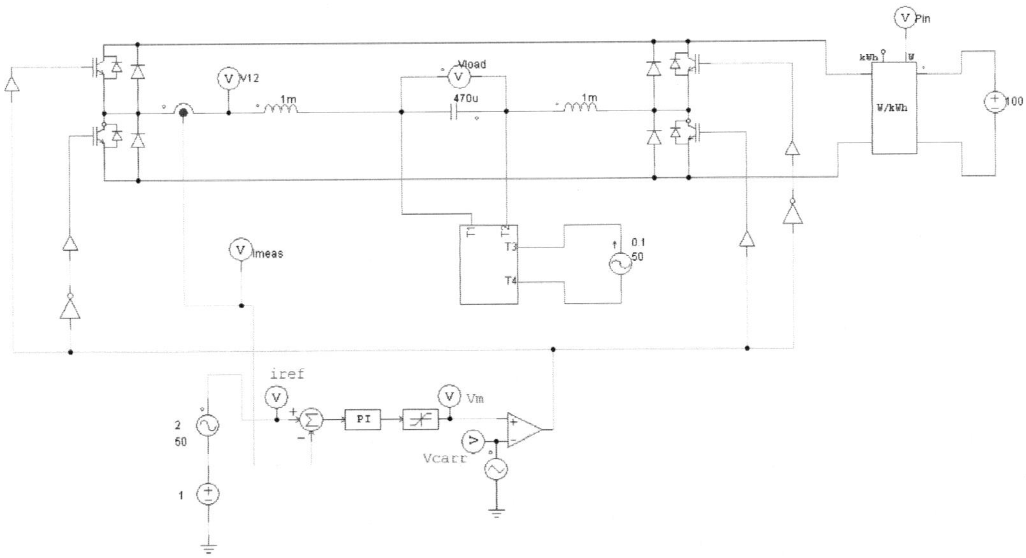

Figure 10a. Transient analysis (PI current controller)

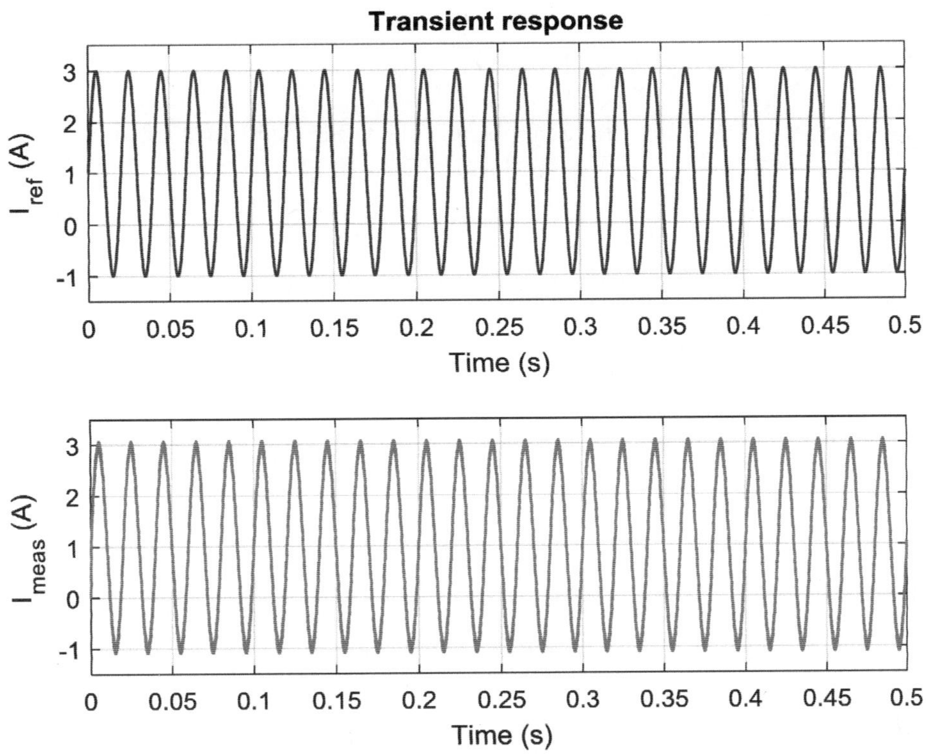

Figure 10b. Transient results (PI current controller). Reference and measured currents.

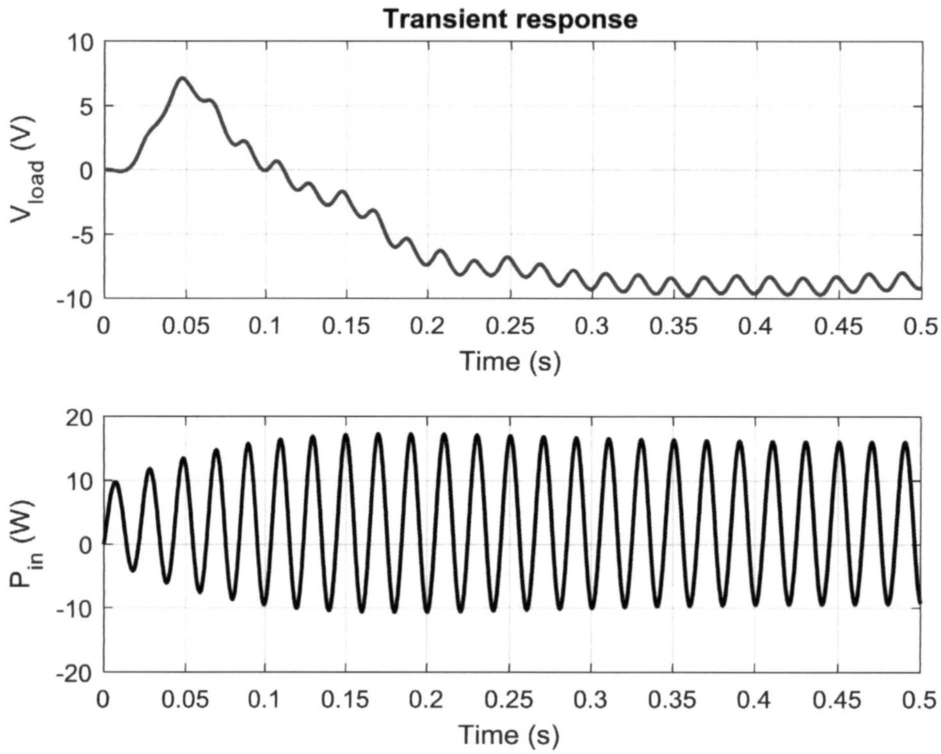

Figure 10c. Transient results (PI current controller). Voltage across the load and power input requirements.

6. Conclusions

The proposed electromechanical analogous circuit is quite useful for studying the dynamics and efficiency of power amplifiers in active vibration control systems. Standard circuit simulation software could be used for tuning the real system.

The proposed current control amplifier shows high efficiency (the losses are largely determined by the ratio of the load resistance to the total DC loop resistance which is the sum of the on resistance of the MOSFETs, wire resistances (including the output filter) and current sense resistor if used, and the load resistance.

The bandwidth of the current control amplifier is quite high which is desired for successfully implement the feedforward controller. The main drawback when compared to the class AB amplifier is the output current ripple. This output current ripple could be minimized increasing the switching frequency and the L value in the output filter.

Simulation results indicate that the proposed current controlled Class D amplifier is an efficient VCM driver for active vibration control.

7. References

[1] Samali, B., Al-Dawod, M., Naghdy, F. *Active Control of Cross Wind Response of 76-Story Tall Building Using a Fuzzy Controller*, Journal of Engineering Mechanics, 130 (4) (2014), pp. 492-498

[2] Perez, A., Hassan, A., Kaczmarczyk, S., Picton, P., Active vibration control using mechanical and electrical analogies, J. Phys.: Conf. Ser. 721 012013

[3] R.F.A. Marques, D.A. Rade, S.S. Cunha, Jr., Assessment of an Active Dynamic Vibration Absorber, IMAC XIX - 19th International Modal Analysis Conference (2001), pp, 32-36.

[4] Cuk, S., Erickson, R.W., A conceptually new-high frequency switched-mode power amplifier technique eliminates current ripple, in Proc. Fifth Nat. Solid-State Power Conversion Conf., (1978), pp. G3.I-G3.22.

[5] Zhang, J., Bidirectional DC-DC Power Converter Design Optimization, Modelling and Control, PhD Dissertation (2008), Virginia Polytechnic Institute

IOP Conf. Series: Journal of Physics: Conf. Series **1048** (2018) 012012 doi:10.1088/1742-6596/1048/1/012012

Vertical vibration of elevator compensating sheave due to brake activation of traction machine

S Watanabe[1] and T Okawa[2]

[1] Advanced Technology R&D Center, Mitsubishi Electric Corp., Hyogo, Japan
[2] Inazawa Works, Mitsubishi Electric Corp., Aichi, Japan

E-mail: Watanabe.Seiji@ay.MitsubishiElectric.co.jp

Abstract. This paper shows an elevator dynamic model that calculates the compensating sheave motion during a brake activation of the traction machine. A simplified mathematical formulation is derived to evaluate the vertical static displacement of the compensating sheave. The vertical vibration induced by the brake of the traction machine is also evaluated and it is concluded that the maximum vibration occurs when the vertical vibration by the car stop timing synchronizes the phase of the vibration by the brake activation. The maximum vibration is also evaluated by a mathematical formulation and it shows that the vertical vibration is proportional to the building height. The derived equations contribute to the elevator's optimal design.

1. Introduction

Most elevators applied to tall buildings include compensating ropes to satisfy the balanced rope tension between the car and the counter weight. The compensating ropes receive tension by the compensating sheave, which is installed at the bottom space of the elevator shaft [1, 2]. The compensating sheave is only suspended by the compensating ropes, therefore, the sheave can move vertically during the car traveling. It is important to evaluate the vertical displacement of the compensating sheave, because the displacement is one of the key factor to determine the pit depth of the elevator shaft.

This paper shows the static displacement and the vertical vibration of the compensating sheave. Firstly, an elevator system model is proposed to evaluate the vertical motion of each component. The derived simulation model indicates that the vertical static displacement depends on the car position and the car loading condition. Based on the simulation results, we can produce a simplified mathematical formulation to evaluate the static displacement.

Secondary, the vertical vibration induced by the brake of the traction machine is evaluated numerically. As the simulation results correspond with the experimental ones, we can investigate the worst condition of the vibration by changing the elevator's system parameters. It is concluded that the maximum vibration occurs when the vertical vibration induced by the car stop timing synchronizes the phase of the vibration induced by the brake activation. By the result, we can also introduce a mathematical formulation to evaluate the maximum vibration.

Finally, we evaluate the relation between the vertical vibration and the building height. By the derived equation, we can conclude that the amplitude of the vertical vibration induced by the brake is proportional to the building height.

In the end, the derived simulation model and mathematical formulations contribute to the elevator's optimal design, especially for the pit depth evaluation.

2. Multi-body dynamic model

Figure 1 shows the multi-body dynamic model of our elevator system. Since the suspension and compensating ropes are treated as continuous bodies, they are modeled by alternating the intensive mass and spring [3, 4].

The multi-body dynamic model in figure 1 is represented by the following equation of motion:

$$\boldsymbol{M}\ddot{\boldsymbol{x}} + \boldsymbol{C}\dot{\boldsymbol{x}} + \boldsymbol{K}\boldsymbol{x} = \boldsymbol{F}, \tag{1}$$

where $\boldsymbol{M}, \boldsymbol{C}, \boldsymbol{K}$ and $\boldsymbol{F}$ are the system's inertia matrix, damping matrix, stiffness matrix and generalized force vector respectively. $\boldsymbol{x}$ is the vector of state variables, which includes the translational and rotational values. Equation (1) includes models of rope slip on the sheaves, tension loss, and the transient brake force of the traction machine [5].

When the brake is activated at the traction machine due to abnormal behaviors in the elevator system, the car and the counter-weight receive a large vertical vibration at the same time. The vibration also induces the vertical motion against the compensating sheave (figure 2).

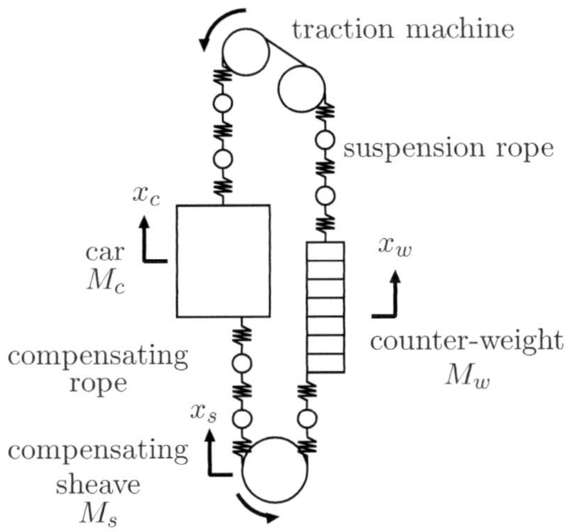

Figure 1. Elevator configuration

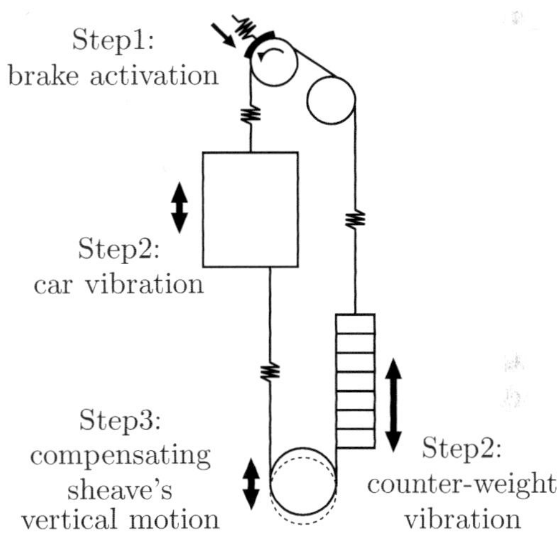

Figure 2. Sheave's vertical motion

3. Simulation results

3.1. Static displacement of the compensating sheave

The vertical displacement of the compensating sheave x_s in figure 1 is evaluated in each car position. Figure 3 shows the compensating sheave's vertical displacement when the car moves downward from the top floor to the bottom floor. During the acceleration and deceleration time, the compensating sheave moves downward. The simulation result of equation (1) shows the same time response against the experimental one.

When the car stops at the top floor, the stiffness of the suspension rope at the car side is high enough. Therefore the car loading condition doesn't affect the compensating sheave's vertical motion at the top floor. The vertical position of the compensating sheave is determined by the stretch of the suspension rope at the counter-weight side. On the other hand, when the car stops at the bottom floor without any load, the sheave moves upward compared to the top floor's stopping condition as shown in figure 3 and figure 4(real line). This is because the stretch of the suspension rope at the car side is shorter due to lighter weight of the car against the counter-weight as shown in figure 5. If the car has a rated load, the sheave moves downward (dashed line in figure 4) due to heavier weight of the car against the counter-weight.

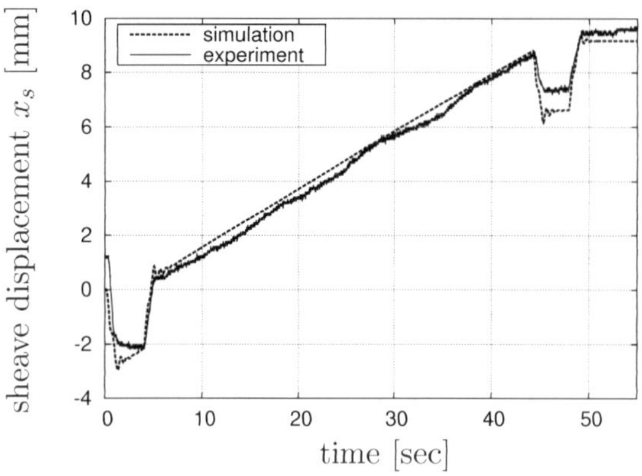

Figure 3. Sheave's vertical displacement x_s during car downward motion (no loading) [3]

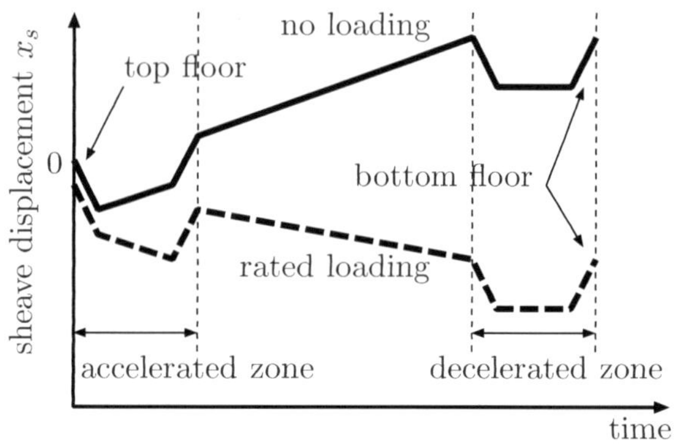

Figure 4. Displacement difference between no loading and rated loading

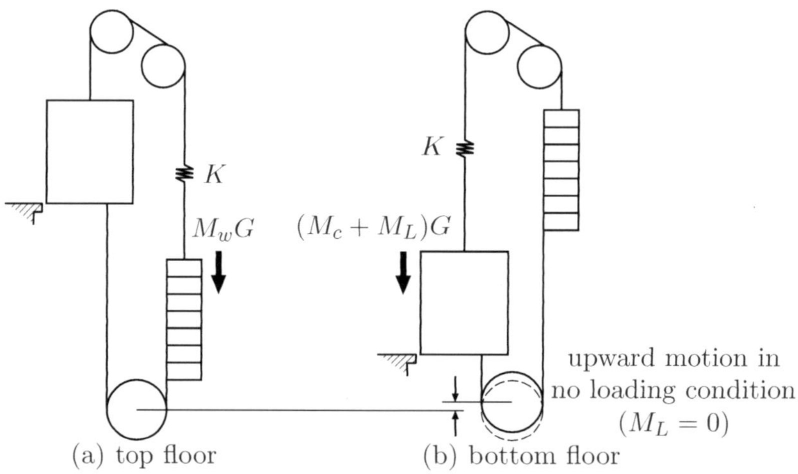

Figure 5. Relation between sheave's vertical displacement and car position

The equivalent rope stiffness K in figure 5 is expressed by the serial spring connection between the suspension rope and the shackle spring at the rope end.

$$K = n \times \frac{K_s K_r}{K_s + K_r}, \tag{2}$$

where K_r and K_s are the stiffness of the suspension rope and the shackle spring respectively. n is the number of rope in the elevator system.

By using the equivalent rope stiffness K, the relative displacement of the compensating sheave x_s is represented by the following simple equation:

$$x_s = \frac{x_d}{2}, \qquad x_d = \frac{M_w G}{K} - \frac{(M_c + M_L)G}{K} = \frac{(M_w - M_c - M_L) \times G}{K}, \tag{3}$$

where G, M_w, M_c and M_L are the gravitational acceleration, the counter-weight mass, the car mass and the car loading mass respectively. x_d is the relative stretch of the suspension rope between the car side and the counter-weight side. As the compensating sheave is the movable pulley, the vertical displacement x_s is half of x_d.

3.2. Vertical vibration during the brake activation

When the brake is activated during the car running, the compensating sheave receives a large vibration as shown in figure 6. The simulation result of equation (1) shows the same time response against the experimental one. The vibration behavior can be described by figure 7.

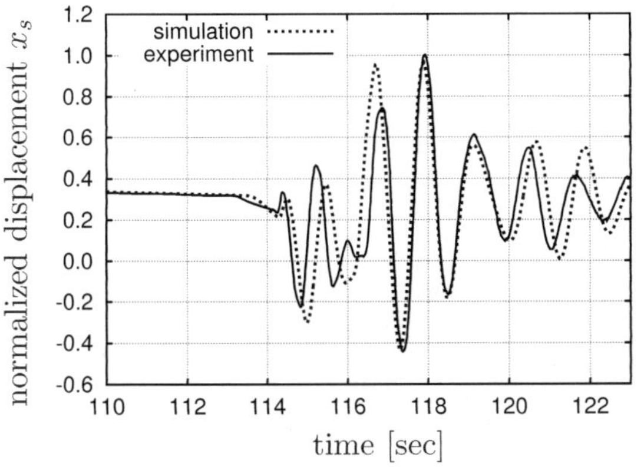

Figure 6. Sheave displacement during brake activation

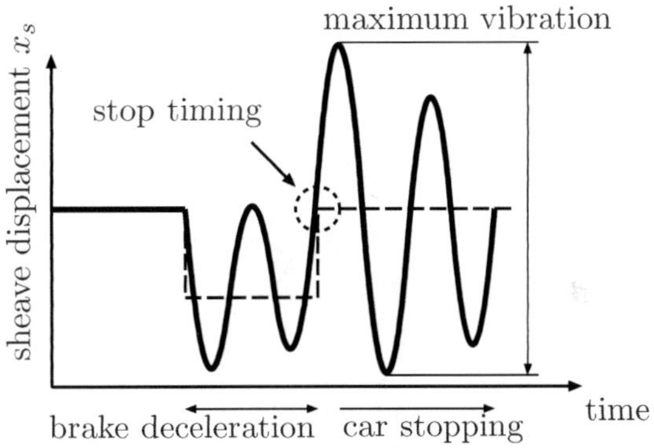

Figure 7. Relationship between sheave vibration and brake timing

The car and the counter-weight receive a step force by the brake. The force induces a vertical vibration and the vibration continues while the car slows down its speed. When the car stops, the compensating sheave receives another step force and then it induces another vibration.

If the first vibration and the second vibration are synchronized at the car stopping time, the second vibration is magnified and a larger vibration occurs after the car stopping time as shown in figure 7.

In this section, we evaluate the above large vibration at the car stopping time. At first, we introduce a generalized mass M, which represents the car weight or the counter-weight one.

$$M = \begin{cases} M_c + M_s/2 + M_r + M_L & \text{(bottom floor condition)} \\ M_w + M_s/2 + M_r & \text{(top floor condition)} \end{cases}, \tag{4}$$

where M_s and M_r are the weight of compensating sheave and suspension rope respectively. The rope weight M_r is expressed as the following equivalent mass.

$$M_r = n \times \rho \times L/3, \tag{5}$$

where ρ and L are the linear density of the suspension rope and the rope length respectively.

As the generalized mass M is suspended by the rope, the mass causes vertical vibration by the brake force of the traction machine. In the actual system, the vertical vibration decreases gradually due to the rope damping effect. However we ignore the damping effect to evaluate the worst condition.

The car velocity pattern also affects the vibration. In figure 7, the brake is activated under the constant speed condition. On the other hand, figure 8 shows the vibration behavior under the brake occurrence at the car acceleration time.

When the car is accelerating at βG, the brake is activated and the car is decelerated to $-\alpha G$ by the brake force.

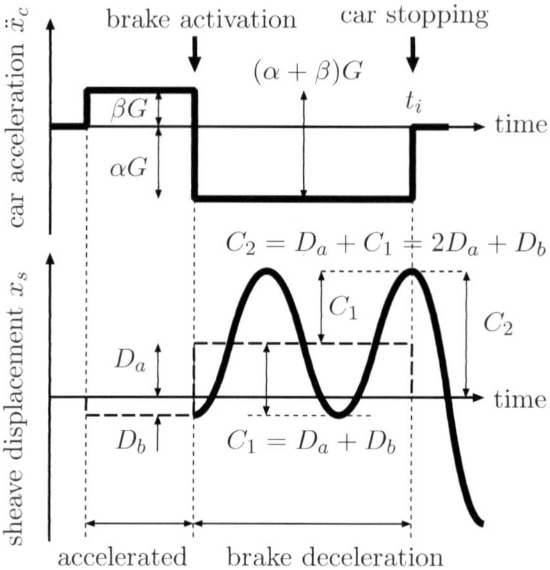

Figure 8. Brake activation at car acceleration time

Due to the time response of the acceleration in figure 8, the vibration of M induced by the brake activation is derived by a simple 1-DOF(degree of freedom) vibration model as shown in figure 5.

$$x_i(t) = \frac{M \times G}{K} \left\{ -\beta + (\alpha+\beta)(1-\cos\omega t) \right\}, \tag{6}$$

where ω is the natural frequency of the generalized mass M suspended by the equivalent rope stiffness K.

$$\omega = \sqrt{\frac{K}{M}}. \tag{7}$$

Equation (6) is derived by the following initial condition.

$$x_i(0) = -\frac{M \times \beta G}{K}, \quad \dot{x}_i(0) = 0. \tag{8}$$

By the step function of the brake force $(\alpha+\beta)G$, equation (6) shows the sinusoidal wave response.

In figure 8, D_a and D_b correspond to the following equation.

$$D_a = \frac{\alpha MG}{K}, \quad D_b = \frac{\beta MG}{K}. \tag{9}$$

From the second term of equation (6), the vibration amplitude C_1 during deceleration due to the traction brake is given by

$$C_1 = D_a + D_b = \frac{(\alpha+\beta)MG}{K}. \tag{10}$$

The residual vibration after the car stopping is also described by a sinusoidal wave of the same frequency in equation (7). The following equation is derived by a simple 1-DOF vibration model in the same manner as equation (6).

$$z(t) = A\cos\omega(t-t_i) + B\sin\omega(t-t_i), \tag{11}$$

where t_i is the time of car stopping and the arbitrary values A, B are determined by the following initial condition.

$$z(t_i) = x_i(t_i), \qquad \dot{z}(t_i) = \dot{x}_i(t_i). \tag{12}$$

Then equation (11) is expressed by

$$z(t) = x_i(t_i)\cos\omega(t-t_i) + \frac{\dot{x}_i(t_i)}{\omega}\sin\omega(t-t_i) = \sqrt{x_i^2(t_i) + \left(\frac{\dot{x}_i(t_i)}{\omega}\right)^2}\sin(\omega t+\phi), \tag{13}$$

where ϕ is the constant of phase shift.

The maximum amplitude of equation (13) is derived by the following condition.

$$\cos\omega t_i = -1 \quad \longrightarrow \quad t_i = \frac{\pi(2N-1)}{\omega}, \quad (N = 1, 2, \cdots). \tag{14}$$

In this case, the maximum amplitude is given by

$$Z = \max\{z(t)\} = \frac{M \times G}{K}(2\alpha + \beta). \tag{15}$$

In figure 8, the maximum amplitude C_2 as the residual vibration is given by the following equation.

$$C_2 = D_a + C_1 = 2D_a + D_b = \frac{(2\alpha+\beta)MG}{K} \equiv Z. \tag{16}$$

The compensating sheave's vibration is half of the generalized mass M, because it moves by the mechanism of movable pulley. Therefore the maximum vibration of the compensating sheave is expressed as follows.

$$x_s = \frac{Z}{2} = \frac{M \times G}{2K}(2\alpha + \beta). \tag{17}$$

3.3. Influence of building height

Equation (17) is the function of the equivalent rope stiffness K. If the rope is long enough, equation (2) is approximated by

$$K \approx n \times K_r = n \times \frac{E_r A_r}{L}, \tag{18}$$

where E_r and A_r are the rope's Young modulus and the cross section area respectively. The rope length L can be approximated by the building height.

Substituting equation (18) into equation (17), it is assumed that the maximum vibration is proportional to the rope length L, or the building height.

$$x_s \approx \frac{MG}{2nE_rA_r}(2\alpha + \beta)L. \tag{19}$$

To evaluate the above assumption, the maximum vibration is calculated in each building height. Figure 9 shows the several simulation results of the maximum vibration in the actual elevator specifications.

The result of equation (17) which is marked by a circle matches the simulation result marked by a square. As both results are proportional to the building height, the above assumption is acceptable.

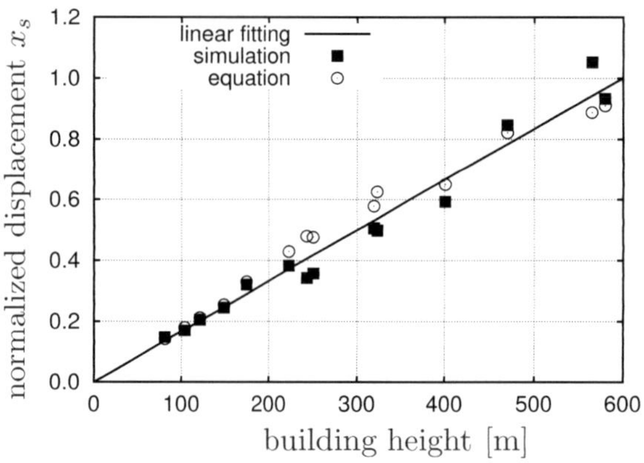

Figure 9. Maximum vertical vibration with building height

4. Conclusions

We evaluated the vertical displacement and vibration of the compensating sheave. Our formula suggests that the static displacement depends on the car position and the car loading condition. We also get another conclusion that the maximum vertical vibration induced by the brake of the traction machine is proportional to the building height. As the maximum vertical vibration and the static displacement are expressed by simple formulas, we can use them for the elevator's optimal design.

References

[1] Crespo R S, Kaczmarczyk S, Picton P, Su H and Jetter M 2013 *Proc. of 3rd Symp. on Lift and Escalator Technologies (Northampton)*
[2] Nakazawa D, Watanabe S, Fukui D and Okawa T 2013 *Proc. of ICOVP-2013 (Lisbon)*
[3] Watanabe S, Okawa T, Nakazawa D and Fukui D 2013 *J. Phys.: Conf. Series* **448** 012007
[4] Watanabe S and Okawa T 2016 *J. Phys.: Conf. Series* **721** 012015
[5] Watanabe S, Yumura T, Hayashi Y and Takigawa Y 2002 *Proc. of ACMD'02 (Fukushima)*

Institute of Physics
Dirac House, Temple Back
Bristol BS1 6BE UK

ISSN: 1742-6588
ISBN 978-1-5108-6789-5
9 781510 867895